The Chamars

THE RELIGIOUS LIFE OF INDIA

EDITED BY

J. N. FARQUHAR, M.A., D.Litt.

LITERARY SECRETARY, NATIONAL COUNCIL, YOUNG MEN'S CHRISTIAN ASSOCIATIONS, INDIA AND CEYLON;

AND

NICOL MACNICOL, M.A., D.Litt.

ALREADY PUBLISHED

THE VILLAGE GODS OF SOUTH INDIA. By the BISHOP OF MADRAS.

THE AḤMADĪYA MOVEMENT. By H. A. WALTER, M.A.

VOLUMES UNDER PREPARATION

THE VAISHṆAVISM OF PAṆḌHARPUR. By NICOL MACNICOL, M.A., D.Litt., Poona.

THE CHAITANYAS. By M. T. KENNEDY, M.A., Calcutta.

THE ŚRĪ-VAISHṆAVAS. By E. C. WORMAN, M.A., Madras.

THE TAMIL ŚAIVA SIDDHĀNTA. By GORDON MATTHEWS, M.A., B.Litt., Coimbatore.

THE VĪRA ŚAIVAS. By W. E. TOMLINSON, Gubbi, Mysore.

THE BRĀHMA MOVEMENT. By MANILAL C. PAREKH, B.A., Rajkot, Kathiawar.

THE RĀMAKṚISHṆA MOVEMENT. By J. N. C. GANGULY, B.A., Calcutta.

THE KHOJAS. By W. M. HUME, B.A., Lahore.

THE MĀLĀS AND MĀDIGĀS. By the BISHOP OF DORNAKAL; P. B. EMMETT, B.A., Kurnool, and S. NICHOLAS, Cuddapah.

THE DHEḌS. By MRS. SINCLAIR STEVENSON, M.A., D.Sc., Rajkot, Kathiawar.

THE MAHĀRS. By A. ROBERTSON, M.A., Poona.

THE BHILS. By D. LEWIS, Jhalod, Panch Mahals.

THE CRIMINAL TRIBES. By O. H. B. STARTE, I.C.S., Bijapur.

EDITORIAL PREFACE

THE purpose of this series of small volumes on the leading forms which religious life has taken in India is to produce really reliable information for the use of all who are seeking the welfare of India. Both editors and writers desire to work in the spirit of the best modern science, looking only for the truth. But, while doing so and seeking to bring to the interpretation of the systems under review such imagination and sympathy as characterize the best study in the domain of religion to-day, they believe they are able to shed on their work fresh light drawn from the close religious intercourse which they have each had with the people who live by the faiths herein described: and their study of the relevant literature has in every instance been largely supplemented by persistent questioning of those likely to be able to give information. In each case the religion described is brought into relation with Christianity. It is believed that all readers in India at least will recognize the value of this practical method of bringing out the salient features of Indian religious life.

VILLAGE TEMPLE IN CHAMAR MAHALLA, DEDICATED TO JAWALI DEVI

THE RELIGIOUS LIFE OF INDIA

THE CHAMĀRS

BY

GEO. W. BRIGGS, M.Sc.

ASSOCIATION PRESS
(Y.M.C.A.)
5 RUSSELL STREET, CALCUTTA

HUMPHREY MILFORD
OXFORD UNIVERSITY PRESS
LONDON, NEW YORK, TORONTO, MELBOURNE,
BOMBAY AND MADRAS
1920

DS
422
C3B7

AUTHOR'S PREFACE

The aim in writing the following pages has been to present an accurate and fairly complete account of the Chamārs. To do so a considerable amount of material has been included which, with variations, is the common possession of many castes. No attempt, however, has been made at a comparative study. The basis of this work has been the Chamars of the United Provinces, but the Chamars and the leather-workers of other parts of India as well have been noted. The writings of Ibbetson, Crooke, Rose, Russell and others have been made use of, and Census Reports, both Imperial and Provincial, have been examined. But apart from facts which could be found only in the Census Tables, nearly all the materials from these sources, which have been incorporated in this book, have been tested in two or more important sections of the Chamars and some matters in other sub-castes as well, and have been verified or modified to fit this particular caste. Similarly, materials from works on anthropology, ethnology, animism, and magic have been made the basis of investigation. In every instance the questions have been, "Is the belief or the practice current among the Chamars?" and "Is this the way the Chamars themselves believe and act?" Men of many sub-castes and of all sorts have been questioned, farmers, tanners, shoemakers, wizards, gurus, and servants. Both the men of the villages and the residents of the towns and cities have been interrogated. The single aim has been in all cases to record the Chamar point of view. The Chamars of the north-west have been influenced by the superstitions of the Punjab, while those to the east reflect the peculiar beliefs of the Vindhyas. On this account uniformity of details in names and in beliefs will not be found. But the fact that certain practices and some names are not traceable in a certain sub-caste or in some locality does not invalidate such matters as Chamar facts. The main outlines of thought and life are, however, fairly uniform throughout the caste.

It has been decided to use diacritical marks in Indian words and to print them in italics only on their first occurrence. In the Chamar sub-caste names, however, diacritical marks have not been used, except in the case of the Chāmar, because it has been impossible to obtain sufficient accuracy of spelling in many instances.

Thanks are due to friends in civil and in missionary circles for help in collecting data and in criticizing the results of investigation.

ALLAHABAD, G. W. B.
January 31, 1920.

CONTENTS

CHAP.		PAGE
	Author's Preface	7
I.	The Caste	11
II.	Social and Economic Life	35
III.	Domestic Customs: Birth	60
IV.	Domestic Customs: Marriage	72
V.	Domestic Customs: Death: Miscellaneous ..	99
VI.	The Spirit World	121
VII.	The Mysterious	158
VIII.	Higher Religion	198
IX.	The Outlook	224
	Appendix A. Tables	248
	Appendix B. Tanning, Shoemaking and Leather Articles	256
	Appendix C. Bibliography	261
	Glossary	263
	Index	266

LIST OF ILLUSTRATIONS

1. Village Temple in Chamār Mahalla .. .. *Frontispiece*
2. Jatiya Chamār—Dhariya Chamār.. .. *Facing page* 22
3. Chāmar Chamār—Nalchina Chamār .. .. ,, 23
4. Pencil Drawings of Shasti and Salona.. .. ,, 66
5. Pencil Drawings of Abdominal Brand Marks.. ,, 67
6. Marriage Pole (Sūgā)—Kohbar.. .. .. ,, 78
7. Pots set in Roof—Blackened Pot with White Spots—Pots around a Bamboo—Chāmuṇḍā's Platform .. .. ,, 79
8. Pot showing Place of Worship—Place of Worship, with Offerings .. .. ,, 146
9. Shrine of Nat Bāba—Kālkī and her Court—Shrine of Hem Raj .. .. ,, 147
10. Pots used as Offerings—Pot and Evil Eye .. ,, 162
11. Pencil Drawings of Magic Symbols .. .. ,, 163
12. Śiv Nārāyan Mahant—Devil Priest .. .. ,, 212
13. Kuril Chamār—Jaiswar Chamār .. .. ,, 213

CHAPTER I

THE CASTE

THE tanners of leather, the preparers of skins, the manufacturers of leather articles, and the makers of shoes belong to a well defined class in the Indian social order. Most of these workers, in Upper India, are to-day included under the general term *Chamār*. This occupational group may be traced back to very early times. Tanners (*carmamnā*) are mentioned in the Ṛig Veda,[1] in the later Vedic literature,[2] and in the Brāhmaṇas.[3] Tanning, *mlā*, *mnā*, is also spoken of in the Ṛig Veda,[4] and certain details of stretching[5] and wetting[6] hides probably refer to the process of manufacture. Ox-hides were used in the pressing of the soma,[7] and ox-hides[8] and antelope[9] and tiger skins[10] were used in sacramental and ceremonial rites. The use of skins for clothing is mentioned in the Śatapatha

[1] VIII. 5, 38.
[2] Vāj. Saṁh. 30, 15.
[3] Tait. Br. III. 4, 13, 1; Ait. Br. V. 32, Carmaṇya (leather work).
[4] VIII. 55, 3.
[5] Śat. Br. II. 1. 1. 9.
[6] R. V. 1. 85. 5.
[7] R. V. X. 94. 9; X. 116. 4.
[8] A. V. XII. 3. Śat.Br. VII. 3. 2. 1-4. Gobhila Gṛh. S II. 3. 3; II. 4. 6; Hiraṇ. Gṛh. S. I. 7. 22. 8; Āpas. Gṛh. S. II. 6. 8; Śāṅkhā. Gṛh. S. I. 16. 1; Āśv Gṛh. S. I. 8. 9; I. 14. 3; IV. 6 4; Pāras. Gṛh. S. I. 8. 10.
[9] A. V. XI. 1. 8; Śat. Br. I. 1. 4. 3; I. 2. 1. 14; I. 9. 2. 33; III. 2. 1. 1-9; III. 3. 4. 1. 8; III. 6. 3. 18; VI. 2. 2. 39; XI. 8. 4. 3; XII. 8.3.3, 9, 21. Skins: Baud. II. 10. 17. 20; III. 1. 11, 18.
[10] A. V. IV. 8. 4; Śat. Br. V. 3. 5. 3, V. 4. 1. 9, 11; V. 4. 2. 6; V. 4. 4. 2.

Brāhmaṇa[1] and in other early literature.[2] The Māruts wore deer skins,[3] and the wild ascetics seem to have been clothed in skins.[4] The use presupposes the preparation of skins.

The word leather (hide) *charman, charma*, is known in both the older and the later portions of the Ṛig Veda,[5] in the Atharva Veda,[6] in the books of the Yajur Vedic schools,[7] in the Brāhmaṇas,[8] and in the later literature.[9] In these books we find references to the thong, *yoktra*,[10] used for yoking the chariot or cart; the bow-string, *jyā*, made of ox-hide[11]; reins of leather; leather bags,[12] *dṛiti* and *dhmāta*, for holding liquids; leather bottles,[13] *bhastrā*; and thongs used for couches, *vardhra*,[14] for door fastenings, *paricarmaṇya*,[15] and for bridles, *syuman*.[16] From the Mahābhārat[17] we learn that leather was used for the hand-guard for the bow; that the hands and fingers were protected with leather; that the soldier used a shield made of ox-hide or of bear-skin: that he had a cuirass and a

[1] V. 2. 1. 21. 24.

[2] Vasish. XI. 61-63; Āpa. I. 1. 2. 40; I. 1. 3. 3, 4, 5, 6, 7, 9; Baud. I. 6. 13. 13; Guat. I. 16; Manu XI. 109; Inst. of Vishṇu, XXVII. 20 (antelope, tiger, and he-goat.)

[3] R. V. I. 166. 10.

[4] R. V. X. 136. 2.

[5] *E. g.*, III. 60. 2; IV. 13. 4; I. 85. 5; I. 110. 8; I. 161. 7

[6] *E. g.*, V. 8. 13; X. 9. 2; XI. 1. 9.

[7] *E. g.*, Tait. Saṁh. III. 1. 7. 1; VI. 1. 9. 2.

[8] *E. g.*, Tait. Br. II. 7. 2. 2.

[9] Vasish. III. 53; Baud. Gṛh. S. I. 1. 1. 10; I. 5. 8. 38, 43; Gaut. Gṛh. S. I. 33.

[10] R. V. III. 33. 13; V. 33, 2; A. V. III. 30. 6; VII. 78. 1; Tait. Saṁh. I. 6. 4. 3; Tait. Br. III. 3. 3. 3; Śat. Br. I. 3. 1. 13; VI. 4. 3. 7. Bṛh. (Minor Law Book) XI. 16.

[11] R. V. VI. 75. 3; A. V. I. 1. 3. There are many passages.

[12] R. V. I. 191. 10; IV. 51. 1. 3; V. 83. 7; VI. 48. 18; VI. 103. 2; VIII. 5. 19; VIII. 9. 18; A. V. VII. 18. 1; Tait. Saṁh. I. 8. 19, 1; Vāj Saṁh. XXVI. 18. 19; Tait. Br. I. 8, 3, 4; Pañc. Br. V. 10. 2. *dṛiti*; R. V. VII. 89. 2, *dhmāta*. These used also as bellows.

[13] Śat. Br. I. I, 2, 7; I. 6. 3. 16.

[14] A. V. XIV, 1, 60; Śat. Br. V. 4. 4. 1.

[15] Kaush. Br. VI 12; Sāṅk. Ār. II. 1.

[16] R. V. III. 61. 4. See also Manu VIII. 292.

[17] "The Social and Military Position of the Ruling Caste in Ancient India as Represented by the Sanskrit Epic." Hopkins in J. A. O. S. Vol. XIII. See especially section IV.

breast-plate of leather; and that his body-armour was made of iron and leather. We find also that sinews were used to bind the feathers upon the arrow, and that the sword was sheathed in leather. The war chariot was protected with shields of leather. The box of the chariot was fixed to the axle with thongs of leather. The horses were yoked to the pole of the chariot with leather straps. The reins were of leather. Sometimes the horses were even covered with leather robes which served as armour. Drums, especially the great kettle-drums, were fashioned with leather heads.

The old literature also knows the shoemaker, *carmakāra, charmakṛit, pādukāra, pādukṛit.* Shoes made of skins and of leather, are mentioned in the Brāhmaṇas,[1] in Manu[2] and the older law books,[3] in the Mahābhārat,[4] in the Rāmāyaṇa,[5] and in the Vishṇu Purāṇa.[6]

Thus there were well known and fully developed in ancient India, the occupations of tanner and leather worker.

Probably from early Aryan times the village life in India was organized somewhat as it is to-day, with its cultivators resident within the village, and the lower orders of labourers attached to its outskirts.[7] To this latter class belonged the common labourers and those who, on account of the disgusting aspects of their work and life, were deemed to be unclean and untouchable. The Aryan came as a conqueror, and he retained for himself the religious and the military functions of the social order, along with the privileges belonging to the leisured class.

[1] Śat. Br. V. 4. 3. 19 (made of boar's skin).

[2] IV. 66, 74. Commentator says, "Colloquially, *jūtā*, leather shoe." See Rājārādhākāntdeva's *Śabdkalpadruma* (Lexicon) under *pādukā*, Vol. II. p. 111.

[3] Āpas. I. 2. 7. 5; Gaut. IX. 5, 45.

[4] *E.g.*, II. 1915. III. 16593; XIII. 4642. See Monier-Williams, *Sanskrit Dictionary*, under "*upānāh*" and "*pādukā*."

[5] *Pādukā*; *jūtā*, leather shoe. Ayodhyākāṇḍa, 112. 29.

[6] II. 21.

[7] See Baden-Powell, *The Origin and Growth of Village Communities in India*, p. 9, note 1. He says that low-caste menials of northern villages are not *part* of the village community. The village community consists of invaders and colonists, the landlords of the village area. See also note 1, next page.

So, as time went on, he became, more and more, the priest and noble, the great landed proprietor and the ranchman. The conquered people, kept in subjection, performed the more lowly tasks of life. According to Hopkins,[1] the *Vaiśya* (the people-caste) and the *Śūdra* (the serving-caste) formed the strata between the ruling and priestly castes on the one hand and the helots (the most depressed classes, the outcastes, the *Dasyu*) on the other. A casual reading of the law books reveals the fact that a fairly sharp line of distinction was drawn between the general community in the village and the helots, who lived beyond the village border.[2] Manu's famous passage is: "All those tribes in this world, which are excluded from (the community of) those born from the mouth, the arms, the thighs, and the feet (of Brahman), are called Dasyus, whether they speak the language of the Mlecchas (barbarians) or that of the Aryas."[3] This excluded group was composed of mixed castes and of aborigines. Some such general term as *Chaṇḍāla* was applied to those who were of polluted Aryan blood, and that of Dasyu (slave, native) to those whom the Aryas had conquered. Sometimes these two words are used as synonyms.[4] The Dasyu was looked upon as inferior and unclean even in Vedic times.[5]

[1] *Essay on Caste.*

[2] There are many passages pointing to this:

(a) Showing different from Śūdra: Āpas. 1. 3. 9. 9, 15; II. 4. 9. 5; Baud. I. 5. 9. 7; I. 5. 11. 36; II. 1. 2. 18; II. 2. 3, 40-42; Inst. Vishṇu III. 32; V. 10, VIII. 2; XVI; XXXV. 3; LI. 11, Gaut. II. 35, "all castes excepting . . . and outcastes." IV. 27, 28; XIV. 30; XV. 24; XXIII. 32. Vasish. XI. 9; XIII. 51; XIV. 2; XV. 13, 17; XVIII. 18; XX. 17; XXIII. 33, 34; Manu III. 239; IV. 79; IV. 213; VIII. 66, 68; XI. 224. Inst. Vishṇu XVI.7-14. "Chaṇḍālas must live out of the town . . ." LVII. 4; LVII. 14. LI. 57; LXXXI. 16, 17. Baudh. II. 3. 6. 22.

(b) Showing that they belong outside the village:

Manu X. 51, "but the dwellings of Chaṇḍālas and Svapachas shall be outside the village"; Inst. of Vishṇu XLIV. 9, calling them untouchable; Inst. Vishṇu LIV. 15, "of Chaṇḍālas and of other low castes that dwell outside the village"; Manu, X. 39, showing that some were excluded from Aryan society.

[3] X. 45. [4] *E.g.*, Manu I. 131.

[5] Baines, *On Certain Features of Social Differentiation in India*. J. R. A. S. 1894, Art. XIX. p.664.

He was never admitted to the Aryan community.[1] Yet these classes had a sort of landed right, and they were useful in times of disease. Acquainted with primitive superstitions, and in many instances being the officiants in magical rites, in exorcism, and in disease transference, they served, in these capacities, even the higher castes.[2] With this community on the outskirts of the village the tanner and leather-worker were grouped.

Occupationally to-day the Chamar corresponds to the charmamna or charmamla and the charmakara of the past.

Some Brahmanical tradition gives the Chamar a respectable ancestry and attributes his out-caste condition to the violation of Aryan laws. According to Manu,[3] the *Kāravāra*, or leather-worker, has the following ancestry.

Karavara	*Niśāda*[4] Father.	*Brāhman* father. Śūdrī mother.
	Vaidehi[5] Mother.	Vaiśya father. *Brāhmaṇī* mother.

Other reports give him a less respectable pedigree, for he is said also to be the offspring of a Chandal woman (one of the most despised of society,[6] having a Brāhmaṇī mother and a Śudra father) by a man of the fisherman caste.[7] And, again, he is said to be the son of a *Mallāh* (boatman) and a Chandal.[8] But evidently none of these traditions account for the Chamar. At most they claim for him a higher birth than seems at all probable.

Much current tradition ascribes to him a good ancestry. For example, men say that, in the beginning, there was but one family of men and they were all of the highest caste. They worked in the fields, and followed other callings. In this family there were four brothers. It so happened that a cow died one day, and the body lay in the yard until evening. Since no one could be found to remove the carcass, the three older brothers agreed that their younger brother should carry away the body, and

[1] *Ibid.* p. 667. [2] *Ibid.* pp. 664, 5. [3] X. 36. [4] Manu X. 8. [5] Manu X. 17. [6] Manu X. 16.

[7] Crooke, *Tribes and Castes of the North-Western Provinces and Oudh*, Vol. II. p. 169.

[8] Elliot *Memoirs, North-Western Provinces of India*, Vol. I. p. 70.

that, afterwards, when he had bathed, they would receive him on the old footing of equality. To this he agreed. After much pulling and hauling, he managed to drag the carcass to the jungle. When he returned from his bath, his brothers refused to receive him, but compelled him to live at a distance from them. He made a great fuss about it, but his complaints were of no avail. They told him that henceforth he was to do the work of a Chamar, that is, to skin the animals that died, and to make leather and implements of leather. The brothers promised to take care of him in return for these services. Thus the Chamar caste arose. It happened on another day that a buffalo died. This Chamar then said to his brothers, "I am not strong enough to remove this carcass." The body lay in the yard until noon, when it so happened that Śiva, who had come down to look after the welfare of men, passed that way. The three brothers complained to him that the Chamar was unable to remove the body of the buffalo. Then the latter appealed to Śiva for help. The great god then said to the brothers, "It is true that your brother cannot, unassisted, remove the carcass. Let one of you step forward and help him." The brothers all protested. Śiva, then commanded the Chamar to collect a pile of refuse (*kūṛā.*) When this was done, Śiva directed him to urinate upon it; and, as he obeyed, straightway, from the heap, a strong man arose. From this man the *Kuril* sub-caste of Chamars sprang.

Another legend, current among the Agarwāla Baniyas, relates that there was once a Rāja who had two daughters, Chāmū and Bāmū, each of whom had a son of great physical powers. One day an elephant died in the Rāja's grounds, and, as he did not wish to cut its body to pieces, he inquired if there was anyone strong enough to carry the carcass away and bury it. Chāmū's son performed the task, whereupon Bāmū's son declared him an out-caste.[1]

[1] Crooke, *Tribes and Castes of the North-Western Provinces and Oudh*, Vol. II. p. 170. For other forms of the same legend see Rose, *A Glossary of Tribes and Castes of the Punjab and the North-West Frontier Province*, Vol. II. p. 148. See also Crooke, *Tribes and Castes of the North-Western Provinces and Oudh*, Vol. I. p. 22,

According to a third legend, five brothers, Brāhmans, while out walking one day, saw the carcass of a cow by the roadside. Four of the brothers passed it by, but the fifth removed the body. Thereupon he was excommunicated by his brothers. His descendants continue to remove the carcasses of cattle.[1]

These traditions, both ancient and modern, do not, however, account for the origin of the Chamar. They merely show how some persons were degraded into the leather-working group. The caste itself had its origin in that occupational class on the borders of the ancient village. This group, essentially non-Aryan, has maintained itself through the centuries in its traditional occupation. But the caste is to-day a very large one, and it would be difficult to account for it merely on the ground that it has been self-propagating. As now constituted, the caste is made up of a heterogeneous group of peoples. This is illustrated, in the first place, by the fact that most of the sub-castes of the Chamars are found in fairly well defined areas, and these may be described as local groups. Furthermore, some sub-caste names, such as *Azamgarhiya*, *Banaudhiya*, *Kalkattiya*, *Ujjaini*, *Saksena*, *Chandariya*, *Guliya*, *Aharwar*, and *Jhusiya*, are specifically local; while other sub-caste names, such as, *Gangapari*, *Purabiya*, *Uttaraha*, and *Dakkhinaha*, point to definite geographical origins. Some of the local groups of Chamars are of recent origin. For example, there were no Chamars in the Gorakhpur District four hundred years ago.[2]

Furthermore, there are good reasons for believing that the caste has received large recruitments from above. This is illustrated by the case of the Gorakhpur Chamars.[3] Again, there are some rather pronounced variations in the features of members of the caste. This may be illustrated from places as widely separated as Ballia and Meerut. It has been noted that many Chamar women have fine features, and that some Chamars have a better cast of features

[1] Crooke, *Tribes and Castes of the North-Western Provinces and Oudh*, Vol. II. p. 170.

[2] *Gorakhpur District Gazetteer*, 1909 p. 94. [3] *Ibid.*, p. 94.

than is at all common in the social level in which they are found. This may be explained in part by illicit relations which Chamar women have had with men of higher castes; and partly by certain social and religious customs that have prevailed extensively, although now traces of the practices are somewhat difficult to discover.[1] But such explanations are not sufficient to account for widespread characteristics of the higher sort. The Jatiya, for example, is of a higher physical type than some other sub-castes and of lighter complexion. The explanation in his case may be that some occupational demand drew Jaṭs into this lower form of work; or, more likely, that some pressure or penalty resulted in their degradation. Some Jatiyas claim to be descendants of Jaṭs, and many of this sub-caste do resemble these taller and fairer complexioned neighbors. Such sections of the caste as possess markedly superior features must be accounted for through conquest. The subjugation of tribe after tribe has been a recurring phenomenon in India. These movements have occurred over wide areas, and over limited portions of the country as well. Local history fully illustrates this fact, and we may picture the flux of rising and falling tribes and clans under repeated foreign and local waves of conquest, and the consequent reconstruction, in more or less detail, of the social distribution of races and clans as a fairly constant process. This means that the fixed status of an occupational group may go hand in hand with the repeated recruitment of the group by those who have been degraded from better positions. In some instances this may mean that certain clans were unable to maintain their identity and prestige with the changing order, and that consequently they have sunk to lower levels. These contentions are borne out by many *got*, or family, and sub-caste names; for example, *Banaudhiya, Ujjaini, Chandhariya, Sarwariya, Kanaujiya, Chauhan, Chandel, Saksena, Sakarwar, Bhadarauriya,* and *Bundela.* These are names of Rajpūt clans, and, as applied to the Chamar, suggest dependency. This may mean also more

[1] See under Marriage; also *Discussions, Representative Council of Missions of the United Provinces*, 1915, p. 7.

or less racial admixture, as in the case of the Jatiya. Sub-caste names such as *Kori* and *Turkiya* point also to the wide range of racial elements in the caste.

On the other hand, there have been large accessions to the caste from below. Got and sub-caste names show that many Chamars have sprung from the *Dom*, the *Kanjar*, the *Habura*, the *Kol*, the *Jaiswar*,[1] and other casteless tribes. This movement of peoples upwards through successive stages is a well-known phenomenon.

The caste, then, has been recruited from numerous sources. Many people and even whole sections of tribes have risen up from the lower levels and entered the caste, and this process is still going on. On the other hand, various political changes have resulted in the subjugation of large groups, who consequently were forced into this lower stratum. Still, the caste is predominantly non-Aryan in character. This is accounted for by the fact that to the basal group, which was of aboriginal origin, large recruitments have been made from below. On the other hand, it may be that environment [2] and food have played a large part in modifying the physical characteristics of those who have been brought into the caste from above. The basal group has always been large enough to assimilate its recruits to its own standards of temper and character. In the Chamar caste, there is a close and historically complete contact with Indian village life running very far back, and to-day it occupies a place in the social and economic order that agrees very well with that held from early times.

Although he does not meet any of the determining tests of Hinduism,[3] the Chamar is a Hindu. In the Census Report for 1901,[4] certain castes which fall below the twice-born were grouped as follows: Those from whose hands Brahmans will take water; those from whose

[1] See Nesfield, *A Brief Review of the Caste System of the North-West Provinces and Oudh*, p. 22.

[2] See *Census, India*, 1911, Vol. I. pt. 1. pp. 383, 384.

[3] See *Census Report, United Provinces*, 1911, pp. 121. 122. He does not usually call himself a Hindu.

[4] See *Census Report, United Provinces*, 1901, pp. 216 ff.

hands some of the higher castes will take water; those from whom the twice-born cannot take water, but who are not untouchable; those whose touch defiles, but who do not eat beef; and those who eat beef and vermin and whose touch defiles. In this last class the Chamar belongs. He occupies an utterly degraded position in the village life, and he is regarded with loathing and disgust by the higher castes. His quarters (*chamrauṭī, chamarwāṛā*) abound in all kinds of abominable filth. His foul mode of living is proverbial. Except when it is absolutely necessary, a clean-living Hindu will not visit his part of the village. The author of *Hindu Castes and Sects* says that the very touch of a Chamar renders it necessary for a good Hindu to bathe with all his clothes on.[1] The Chamar's very name connects him with the carcasses of cattle. Besides, he not only removes the skins from the cattle that have died, but also he eats the flesh. The defilement and degradation resulting from these acts are insurmountable. The fact that the Chamar is habitually associated in thought with these practices may partially explain why the large non-leather-working sections of the caste are still rated as untouchable.

Chamars, including Mochis, are scattered well over the "Aryo-Dravidian" tract, and leather-workers, under one name or another, are found in nearly every part of India. Chamars are most numerous in the United Provinces, and in the bordering areas of Bihar on the East and of the Punjab on the north-west. The census figures for 1911, for all India, show the Brahmans as the first caste in point of numbers, and the leather-workers as a whole, or even the Chamar-Chambhar taken alone, as the second. The Rajput is the third caste. This estimate excludes in calculation the Sheikh Mussulmans, who number 32,131,342 and who are evidently not a "caste." In Bengal the Chamar-Mochi is the sixth caste, the Brahman being the second, and the Kayastha the third; in Bihar and Orissa the Chamar is the eighth or seventh,

[1] P. 267. It may be of interest to know that in Baluch Mochis and Chamars are classed as Jats. See Risley, *Peoples of India*, p. 121.

according as he is counted alone, or with the Mochi; in the Central Provinces he is the third caste; in the Central India Agency the second, with the Brahman first; in the Punjab he is the fourth, or the third if the Mochi be counted, while the Jat is the first and the Rajput second; in Rajaputana he is third, with the Jat first and the Brahman second; in the United Provinces he is the first caste in point of numbers, with the Brahman second. Another striking fact is that in the United Provinces the Chamars are almost as numerous as the Mussulmans. Furthermore, the Chamar is increasing in numbers. In the United Provinces, during the twenty years ending in 1901, the increase was nearly ten per cent.; and during the last decade, 2.4 per cent. In the last thirty years the increase has been 12.2 per cent.[1]

The tables[2] show that the Chamars are scattered fairly evenly over the United Provinces. Numerically they are strongest in the Gorakhpur and Basti Districts; but, taken in proportion to the rest of the population, they are the largest element in the community in Saharanpur and in the remainder of the Meerut Division. In the Saharanpur District every fifth man is a Chamar, while in the Meerut Division seventeen per cent. of the population are Chamars. Taking the United Provinces as a whole, every eighth man is a Chamar.

The sub-castes[3] of the Chamar are very numerous, 1,156 being returned in 1891.[4] While these returns may not be accurate, and while numerous names are but variable pronunciations and spellings of others, still the number of sub-divisions of Chamars is very large. Like

[1] Hindu Chamars, 1911—6,076,000; 1901—5,932,000; 1891—5,854,000; 1881—5,413,000; The figures for Mussulman and Ārya Chamars are not given. Chamar Sikhs numbered 118,000 in 1911, as over against 260,000 in 1891. The figures are in 1,000's only. *Census Report, United Provinces*, 1911, pp. 376-377.

[2] See Appendix A.

[3] This section on the sub-castes is based upon the works of Risley, Sherring, Ibbetson, Crooke, Rose, Russell, and others, and upon independent investigations.

[4] Crooke, *Tribes and Castes of the North Western Provinces and Oudh*, Vol. II. article "Chamar."

many other castes they are said to be divided into seven principal sub-castes. The names of these traditional seven vary in different places and their order of respectability varies also.

Among all the sections of the Chamar of the United Provinces, two great sub-castes predominate. These are the JATIYA and the JAISWAR. The former, which includes more than twenty per cent. of the total Chamar population, is found almost entirely in the north and west of the Provinces, in the Meerut, Agra and Rohilkhand Divisions, being most numerous in Meerut, Agra, Moradabad, and Badaun Districts; and the latter, numbering about one million persons, are found chiefly in the Allahabad, Benares, Gorakhpur, and Fyzabad Divisions, being most numerous in the Jaunpur, Azamgarh, Mirzapur, and Fyzabad Districts. These two sub-castes make up nearly two-fifths of the whole Chamar population. Both make claims to superior standing; and the Jatiya can reasonably claim to be the highest of all the sub-castes of the Chamars. Among them there are many who are well-to-do. The Jaiswar makes claims to superiority, and bases them upon his refusal to do certain degrading tasks that usually fall to the lot of the Chamar. Yet, where they are most numerous, they undoubtedly share in all of the degrading work, and practise all the disgusting habits characteristic of the caste.

The Jatiya, or Jatua, is found in large numbers, not only in the central and upper Doab, and in Rohilkhand, but also in the Punjab in the neighbourhood of Delhi and Gurgaon. He is a field-labourer, a cultivator, a dealer in hides, and a maker of shoes. Some of the cultivating sections of this sub-caste do not make leather, and do not allow their women to practise midwifery. Some of the shoemaking sections do not mend shoes. In some places, notably in the Punjab, the Jatiya works in horse and camel hides, and refuses to touch the skins of cattle.[1] Some of the dealers in hides are wealthy, and live as comfortably as do high-caste Hindus. About one half of the

[1] Rose, *A Glossary of the Tribes and Castes of the Punjab and North-West Frontier Province*, Vol. II. p. 149.

JATIYA CHAMAR

DHARIYA CHAMAR, ONE OF THE SMALLER GROUPS SO NUMEROUS IN THE LOWER JUMNA VALLEY

CHĀMAR CHAMĀR

NALCHINA CHAMAR

sub-caste eat carrion. Some, at least, refuse to eat beef or pork.

Two suggestions have been made as to their origin. Some say that their name is derived from the word *jaṭ*, meaning a camel-driver; others, that their name connects them with the Jāt caste. It is sometimes said that they are descendants from the marriages of Jāts with Chāmars. Nesfield suggests that they may be an occupational offshoot from the Yadu tribe from which Kṛishṇa came. Although the Jatiya of the Punjab works in camel and horse hides, which is an abomination to the Chandar,[1] he employs Gaur Brahmans, and is, for this reason, in that part of India, considered the highest sub-caste of Chamars.

The Jaiswar is found almost exclusively in the eastern part of the Provinces. From his ranks many menial servants and house-servants for Europeans are recruited in the towns and cities. Many are grasscuts and grooms; indeed many of the grooms (*sāīs*) from Calcutta to Peshawar are Jaiswars of Jaunpur and Azamgarh. Some of this sub-caste are tanners, some of them make shoes, and many are day-labourers. Some Jaiswars were with the troops that fought with Clive at Plassey. It is said that they have a custom which requires that they, because of an oath in the name of the goddess Māī Rām (Kālī), carry burdens on their heads but not on their shoulders. They worship the halter as a fetish, and consider it an act of sacrilege to tie up a dog with it, because the dog is unclean. For the most part they eat carrion and pork, but their leading men do not. In some places Jaiswar women practise midwifery.

The details of certain other important sub-castes of the Chamars, as found in the United Provinces, together with supplementary notes bearing on other areas, are given below.[2] Of these sub-castes the more important have been chosen in the order of their numerical strength.

[1] See page 28.

[2] These figures have been based upon Crooke's notes on the Census of 1891. No later data are available. See his *Tribes aud Castes of the North-Western Provinces and Oudh*, table at end of article "Chamar."

The numbers in the first eight sub-castes enumerated range from more than 400,000 to just under 100,000.

The CHĀMAR Chamar is found almost exclusively in the Meerut and Rohilkhand Divisions. He is most numerous in the Saharanpur, Bijnor, and Muzaffarnagar Districts, and he is found in considerable numbers in the Meerut, Moradabad, and Bulandshar Districts. He is counted amongst the lowest of all the sub-castes. In fact the tanning sections of the Chamars, of whom the Chāmar is one, seem to occupy the lowest level wherever they are found. He is a cultivator, a shoe-maker, and a tanner. His women practise midwifery. He eats pork.

The DOHAR is a numerous group of the Chamars, found in a section running right across the Provinces, from the Districts of Philibhit and Kheri, through those of Shahjahanpur, Hardoi, Farrukhabad, Cawnpore, and Etawah, to Jalaun. He is most numerous in the Hardoi District, where he forms more than half of the Chamar population. He does not keep pigs, but he eats pork.

The KURIL is found chiefly in the Allahabad and Lucknow Divisions. He is most numerous in the Unao District where he comprises nearly the whole of the Chamar community. He is found in considerable numbers in the neighbouring Districts of Cawnpore, Lucknow, and Rae Bareilly, and in small numbers in nearly every district in the Provinces, being in this respect, with the exception of the Jaiswar, the most widely distributed sub-caste in the Provinces. He claims to have been brought to Lucknow from Fatehpur Haswa several generations ago. He is a leather-worker and field-labourer. He keeps pigs and eats carrion. He will not touch dead camels or horses. The Kurils who live to the west of the Ganges have no social intercourse with those who live on the other side of that stream. The two sections do not intermarry. The women of the former wear skirts and those of the latter wear loin-cloths (*dhotī*).

The PURBIYA numbers nearly 300,000. The name is geographical. He is found chiefly in the Sitapur and Kheri Districts, being most numerous in the former.

There are fairly large numbers of this sub-caste in the territory lying to the east of these districts. Few are found in the western parts of the Provinces.

The KORI or KOLI Chamar is found almost exclusively in the Gorakhpur and Lucknow Divisions. About 100,000 are found in the Sultanpur District alone, while more than 50,000 are found in the District of Basti, and more than 80,000 in the two Districts of Fyzabad and Partabgarh. He is a shoe-maker, a field-labourer, a groom, and a weaver.[1] He will not touch dead camels or horses. In the Punjab, where he does not work in leather, and where he does not perform menial tasks, he is called a Chamar-Julaha, *i.e.*, Chamar-weaver. The Kori (Weaver) often lives alongside of him, and was undoubtedly formerly a Chamar. In some places people still remember when the Kori and the Kori Chamar ate together and intermarried. In Mirzapur the Kori is known as Chamar-Kori.

The AHARWAR is found chiefly in Bundelkhand, where in some districts, as in Jhansi and in Hamirpur, he comprises about ninety per cent. of the Chamar population. There are important communities of Aharwars in the Districts of Farukhabad, Hardoi, and Bulandshahr. In some places, he does not make leather, nor does his wife practise midwifery. Many Aharwars are cultivators, and some are petty contractors.

The DHUSIYA or JHUSIYA is found almost exclusively in the Benares Division and in the adjoining District of Gorakhpur. He is most numerous in the District of Ballia where he forms about sixty-five per cent. of the Chamar population. Nearly forty-five per cent. of this sub-caste are found in the Ballia District alone. The only other Districts where he is found in considerable numbers are Benares and Gorakhpur. In the Ballia and Benares Districts are found nearly three quarters of the whole sub-caste. Colonies of Dhusiyas are found in the Districts of Saharanpur and Bulandshahr and there are large settlements of them in the Punjab. Although he

[1] Sherring's *Tribes and Castes*, Vol. I. p. 393. See also Elliot, *Memoirs, North-Western Provinces of India*, Vol. I. 70.

is a shoe and harness maker, he is chiefly a day labourer. Some of the sub-caste are tanners. He sometimes serves as a musician. House-servants of Europeans are often from this sub-caste. Occasionally he cultivates his own fields. In the east, *e.g.*, in Bihar, he keeps pigs and chickens. His women practise midwifery. In the Punjab he is counted as a sub-division of the Mochi.

The Chamkatiya is found chiefly in the Bareilly District, where nearly eighty per cent. of the sub-caste is found. There are a few thousands, all told, found in a section running through the Districts of Fatehpur, Rae-Bareilly, Sultanpur, Fyzabad, and Basti. Chamkatiyas are scarcely found elsewhere. It is said that from this sub-caste both Nona Chamārī and Rai Dās came.

The Dosadh or Dusadh, found in the Lucknow and Gorakhpur Divisions and in the lower Doab, is a weaver, a groom, and a field-labourer. He keeps pigs. In Bengal the Dosadh claims to be of higher standing than the Chamar. Formerly, in the east, he was reckoned as a Chamar, but now he assumes an independent position. He no longer works in leather, nor does he eat carrion, nor does his wife practise midwifery. He often works as a house-servant. He is on very friendly terms with the Chamars and lives next to them in the villages. Many Dosadhs have gone to the cities to work in the factories.

From the Azamgarhya, or Birhiruya, of the Gorakhpur Division, come many servants of Europeans. They also tend swine.

The Kaiyan of Bundelkhand and Sagar is sometimes rated as a criminal. He is related to the Bohra, a trader and usurer of Brahman, or Rajput, origin.

There are some groups of Chamars that are often spoken of as sub-castes, which are not strictly such. The Rangiya is a good example. It is an occupational division of certain sub-castes. As the name suggests, he is a dyer, or tanner, of leather, and, as such, is a low type of Chamar. Some of them make shoes. Another group that is often spoken of as a sub-caste is the Rai Dasi. With the possible exception of those in the Karnal and its neighbourhood, this group is not a sub-caste. In some parts of the provinces all Chamars call themselves Rai Dasis, and many bearing this name are found as religious groups in a number of sub-castes. Followers of Rai Dās are found all over the provinces. There are other religious bodies amongst the Chamars which are not sub-castes.

On the other hand the SATNAMIS, a religious group in the Central Provinces, have become practically a new sub-caste. These Chamars, who make up the largest and oldest Chamar group in this part of India, have given up leather work entirely, and have become cultivators. Many of them have tenant rights, and a number of them have obtained villages. Likewise the ALAKHGIR, a group formed by Lālgīr, has become a separate sub-caste.

While it is unnecessary to name all the sub-castes of the Chamars, a number of groups may be added to those already enumerated. The MANGATIYA is a beggar who receives alms from the Jaiswars only. Once a year he makes his rounds, taking a pice and a *roṭi* from each house. The CHANDAUR makes but does not mend shoes, and sews canvas and coarse cloth. The NONA Chamar is found in the neighbourhood of Cawnpore. The DHENGAR and the NIKHAR, tribes of the Etawah District, are Chamars. The former serves as a groom, but the latter does not. Their wives do not practise midwifery. The SAKARWARS are tanners, shoe-makers and cultivators. They keep pigs. The KAROL is a small tribe of shoemakers found in the Bahraich, Aligarh, Bulandshahr, and Benares Districts. Then, there are the DHUMAN, DOMAR, RAJ KUMARI, NIGATI, DHINGARIYA, GHORCHARHA, and PACHHWAHAN. Among the minor sub-castes may be noted the GOLE of Etawah; the DOLIDHAUWA, or palanquin-bearer of Partabgarh; the DHUNYAL-JULAHA, who makes cloth; the LASHKARIYA, who makes shoes, often of the English style, and the GHARAMI, of Dehra Dun, who is a thatcher. The RAJ or RAJ-MISTRI, found everywhere in the United Provinces, a purely occupational caste of masons and bricklayers, is largely recruited from the Chamars. This caste is of comparatively recent origin. The CHAIN, who is in some areas, *e.g.*, in Ballia, rated as a Chamar, is also considered a separate caste.

He is described as a criminal, a thief, a swindler, an impostor, and a pick-pocket. He is decidedly the criminal among the Chamars, making long expeditions with the object of looting and robbing. He is a terror to law-abiding citizens and a thorn in the flesh of the police. He is often under police supervision. The DHANUK is sometimes classed as a Chamar.[1] He eats carrion and the leavings of food from other castes, and his women act as midwives.

There are a number of minor castes that work in leather. The DAFALI makes the drums called *tablā* and *tāśa*, and the BHAND, or jester, makes the drums called *ḍamkā*. There are also the DHOR, who makes buckets and dyes leather; the KALAN, who cobbles shoes and makes tents; the DABGAR, found in Bengal and in the east of the United Provinces, as well as in the Punjab, who makes large raw-hide vessels, beaten raw camel's hide bottles for ghee and oil, and also drum-heads, leather sheaths for swords, and shields; the DHALGAR, a maker of leather shields; the CHAKKILIYAN, the Dom of the hill tracts, and the KORAL are also workers in leather. The KHATIK makes drum-heads. The CHARKATA is a Mohammadan leather-worker. The bihiśti, who is sometimes a Chamar, also works in leather. The CHIK, CHIKWA, is a Mohammadan who turns out goat and sheep skins.

In the Punjab[2] still other sub-castes of Chamars are found. The CHANDAR, whose origin is traced to Benares, is sometimes reckoned as the highest of the sub-castes. He does no tanning. He forms the principal sub-caste in the Hisar and Sirsa regions. The CHAMRANG is a tanner who works in ox and buffalo hides only, and who does not work up the leather which he tans. One section of this group, which keeps pigs, is separated from the other, which dyes and tans hides. The RAMDASI is a weaver. The CHAMBAR is the principal sub-caste about Jalandar and Ludhiana. Besides, there are the CHĀMAR, the CHAMARWA, CHANWAR, and the JATA. The last is the descendant of the

[1] Elliot, *Memoirs, North-West Provinces of India*, vol. I. p. 78.

[2] See Rose, *A Glossary of Tribes and Castes of the Punjab and North-West Frontier Province*, article "Chamār," and Ibbetson, *Punjab Census Report of 1881.*

wife of Rām Dās. In Patiala we have the endogamous BAGRI and DESI. The former is an immigrant from Bagar, and the latter consists of two groups, Chamars who make shoes, and the BONAS, weavers of blankets, who are Sikhs. Among the allied castes in the Punjab are the DHED, who is a separate caste in the Central Provinces, and in Gujarat; but who does there much that is really Chamar work; the BUNIYA and the RUHTIYA, both Sikh Chamars, who have taken to weaving; the BILAI (known as a Chamar in the Punjab), a groom and a village messenger, and, in the Central Doab, a weaver and labourer; the DOSADH, an eastern tribe of Chamars; the RAMDASI, or SIKH, who is usually a weaver, and who does not eat carrion; and the KHATIK. Besides these there is the MOCHI, who is, for the most part, a Mussulman Chamar. He works in leather, graining it and giving it a surface stain. In the west he is a worker in leather, whether it be as a skinner, as a tanner, or as a shoemaker. The name mochi is often applied to the more skilled workman of the towns and cities. The Mochi is not usually a weaver. In the west he does not occupy as important a place in agriculture as in the east. He does not render menial services. Where the Chamar is not numerous, his place is taken by the Mochi. The KHATIK, the PASI, and the CHANAL are traditionally connected with leather worker. The latter is a professional skinner in the Simla hills and corresponds to the Chamar of the plains.[1]

In Behar and Bengal the MOCHI and the CHAMAR are one caste.

In the Central Provinces[2] we have the CHAMARS, the greater portion of whom are in the Chhattisgarh Division. Here many villages contain none but Chamars, from the landlord down; and seventy per cent. of these Chamars have given up leather work entirely. Among the subcastes in these Provinces, the SATNAMI is the most important. Other Chamars are termed *paikahā* as opposed to the Satnami. The KANAUJIYA and the AHARWAR are tan-

[1] See *Census Report of the Punjab*, 1911, pp. 398, 469.
[2] See *Census Report, Central Provinces*, 1911, p. 231.

ners and leather workers. They make shoes in a peculiar way. The Kanaujiya eats pork but does not raise pigs. The Aharwar claims to be a descendant of Rai Dās. The JAISWAR is a groom. There are a number of territorial groups whose names have geographical significance, among whom are the BUNDELKHANDI, the BHADORIYA, the ANTERVEDI, the GANGAPĀRI, the PARDESHI, DESA, or DESWAR, MAHOBIĀ, KHAIJRAHA, LADSE for LADVI, MARATHI, PARVĀRHIYA, BERARIĀ and DAKHINI. There are also a number of groups whose names are of occupational significance. These are the BUDALGIRS, makers of leather bags (*budla*); the DAIJANIYAS whose women folk are midwives (*dāī*); the KATUAS, or leather-cutters; the GOBARDHUAS, who collect the droppings of cattle on the threshing floors, and wash out and eat the undigested grain; the MOCHI, or shoe-maker; and the JINGAR, the saddle-maker and book-binder. The Jingar claims to be superior to the Mochi, although the latter claims to be of Rajput origin; and some under the name, JIRAYĀT, are separating from the main caste and are forming a higher social group. They are skilled artisans who handle guns and other delicate instruments. At the other extreme of the social scale is the DOHAR, who is a grass-cutter and doer of odd jobs. Besides, there is the aboriginal worker in leather, the SOLHA, a very small group. The KOR-CHAMARS are weavers. In Berar we find the superior ROMYA or HARALYA Chamar. Two groups of beggars are the MANGYA and the NONA Chamars. In Raipur the Chamars have become regular cattle-dealers and are known as KOCHIAS. In Central India we find the BALAHIS, one section of whom are weavers, and the other, carrion-eaters, who skin animals and deal in skins. (In the Punjab the Chamars engaged to manure the fields and some who take up groom's work are called Balahis or Balais.)

In the eastern parts of Rajputana, the leather-worker is a Mohammadan. Other leather-workers of this area are BANUBHI, BOLA, MEGHWAL, RAIGAR, JATIYA, CHANDOR, SUKARIYA, MOWANPURIYA, KAUSOTIYA, and DAMARIYA.[1] In Bikaneer the BALAI is the leather-worker.

[1] See *Census Report, Rajputana*, 1901, p. 147.

In the Bombay[1] Presidency are found, as in North India, seven main divisions of leather-workers. Of these, the SATRANGAR and the HALALBHAKT are dyers of skins, the former working in sheepskins; the PARADOSH-PARDESI manufactures tents; and the DABALI, the WOJI, and the CHAUR are lower in the social scale than the others, and eat the flesh of bullocks and of other animals. Besides these, there is the MARATHI CHAMAR and the KALPA. All of these, except the Paradoshpardesi, are shoemakers. There is also the JINGAR, or saddle and harness maker, and the RANGARI, or tanner. In addition to these we have the DHOR, a maker of leather buckets and a dyer of skins; the KATAI, a cobbler and tent-maker; and the DAPHGAR, a bottle-maker. The two last-named eat carrion. In Gujarat[2] we find the KALPA, a skinner and tanner, and the MOCHI, a maker of leather and of shoes.

The leather-worker of the Tamil country is the CHAKALLIYAN.[3] He is a dresser of leather and a maker of slippers, harness, and other articles of leather. He is a devil-worshipper. He holds sacred the *avaram* (*cassia aureculata*) tree. It is to be noted that the bark of this tree is a most valuable tanning agent. The men of this caste are drunkards. They eat flesh, and are more detested than the Pariah. As a usual thing their girls are not married before puberty. Widows are re-married. Divorce is common and is easily secured. Their women are beautiful, and from amongst them is usually chosen the woman for the coarser form of *śakti* worship. The women are noted also for their intrigues with landlords and other rich men.

The great leather-working caste of the Telugu country is the MADIGA.[4] He lives on the outskirts of the village. He is described as coarse and filthy, as an eater of unclean food, and as a user of obscene language. He works in leather, and serves as a menial and as a scavenger. Many

[1] Sherring, *Tribes and Castes*, Vol. II. pp. 203 ff.
[2] *Ibid.* Vol. II. p. 279.
[3] *Castes and Tribes of Southern India*, E. Thurston, Vol. II. pp. 2 ff.
[4] *Ibid.*, Vol. IV. pp. 292 ff. See also General Index, *The Village Gods of South India*, Bishop Whitehead, under "Madigas."

MADIGAS are practically serfs. Most of them are field-labourers. They beat drums at festivals. In some parts of the country they still have their perquisites (*jajmān*), but these are disappearing under competition. They perform the revolting parts of bloody sacrifices, and aid in removing the demons of disease. Their girls are often dedicated to temple service (*basavīs*). The caste is divided into a number of endogamous divisions with exogamous septs, some of which seem to be totemistic. Widows are re-married. Divorce is easily secured. They have a *pañchāyat*, or council. They both bury and burn their dead. In 1902 ten per cent. of the Madigas were returned as Christians.

Evidences of affiliations with other castes have already been mentioned, such as the Kaiyan with the Bohra from above; and the Kori and the Kol and other alliances from below.[1] Other cases of affiliations and illustrations of caste fissure are suggested by such well-known names as, Kori-Chamar, a weaver become tanner; Chamar-Julaha, a Chamar become weaver; and Chamar-Kori. In Gorakhpur there are no Koris, but Kori-Chamars. The KARWAL, a vagrant tribe, is found also as a sub-caste under the Chamar.[2] The Darzi, the Banjara, the Barhai and the Sonar each have a Chamar sub-caste.[3] The Kayastha-Mochi, who makes saddles and harness, claims to be of superior origin, and says that the term, "Mochi" refers merely to his occupation. There are other sub-castes of Chamars and allied castes which now form more or less separate bodies and claim to be distinct castes. Even the Jaiswar, for example, claims, in some places, to be a separate caste. The Dusadhs of Bihar are another example. The Kori (Hindu weaver) is probably another instance of caste fissure.

A notable example of a caste formed from the Chamar is the Mohammadan weaver, the Julaha. He is distributed over the United Provinces in considerable numbers. and is

[1] See pages 25 and 26.

[2] *Census Report, United Provinces*, 1911, p. 368.

[3] See Crooke, *Ethnographical Handbook of The North-West Provinces and Oudh*. pp. 188, 65, 23, 70. See also Russell, *Tribes and Castes of the Central Provinces*, Vol. I. p. 353.

found also in other parts of India, especially in the Punjab.[1] He is a typical illustration of how a group of people may rise in the social scale within the Brahmanic system. Originally a Chamar, he secured a better position by taking to weaving. He eats no carrion, touches no carcasses, does not work in impure leather, and has separated himself entirely from the other sections of the Chamar. In taking to the comparatively high occupation of weaving, he has reached the border of the respectable artisan class. In many places this separation took place a good while ago; but Ibbetson reported instances of this process still going on. His numbers are recruited from several groups, as the following names show: Chamar-Julaha, Koli-Julaha, Mohammadan-Julaha and Rai Das-Julaha. In many instances now the caste prefix has been dropped. Ninety-two per cent. of the Julahas are Mohammadans. Among the Hindu Julahas are many Kabir-panthis and Ramdasis. Kabir was a Julaha.

Still more important is the Mochi, a purely occupational off-shoot from the Chamar. The word "mochi," which is applied to those who make shoes, leather aprons, buckets, harness, portmanteaux, etc., denotes occupation rather than caste. Mochis are divided into two main classes, those who make and cobble shoes, who are real Chamars; and those who make saddles and harness. These latter call themselves *Sirbāstab-Kāyasths*, with whom they intermarry and agree in manners and customs. According to a text cited as authoritative by the pandits of Bengal, the astrologers are shoe-makers by caste, and good Brahmans sometimes refuse to take even a drink of water from their hands.[2] In 1891 there were reported one hundred and fifty sub-divisions of Hindu Mochis.[3] In some places the Mochis of the towns are divided into

[1] See Rose, *A Glossary of Tribes and Castes of the Punjab and the North West-Frontier Province*, Vol. II. 413 ff., and Crooke, *An Ethnographical Handbook for the North-West Provinces and Oudh*, pp. 97,98.

[2] Bhattacharji, *Hindu Castes and Sects*, p. 173.

[3] Crooke, *Tribes and Castes of the North-Western Provinces and Oudh*, Vol. III. p. 498.

functional sub-castes, such as saddlers, embroiderers of saddle-cloth, ghi bucket-makers, makers of spangles and of shields and scabbards. These sub-castes rise in rank as their calling requires greater skill or more costly materials.[1] While the Mochi is an offshoot from the Chamar, as a caste he is quite distinct. However, this holds good in certain areas only. He neither eats nor intermarries with the Chamar. The Mochi does not eat carrion or pork, and his wife does not serve as a midwife. His touch is not polluting. The maker of leather is considered lower in the scale than he who works in prepared leather. As a class he is well off, and socially superior to the Chamar.[2] The Gorakhpur Mochi has received medals at Melbourne and Paris for embossed deerskins, made up as table-cloths, table-mats, carpets, etc. The Bengali Mochi is a Chamar, but he tans only cow, buffalo, goat, and deer hides. Many Mochis are Mohammadans. The Census of 1891 returned twenty-seven divisions of Mohammadan Mochis.[3] The Mochi of Garhwal is from the non-Aryan race called the Dom and is an endogamous group; and in Almora this group includes Chandal (Chamar), and Mochi or *Sarki* (tanner).[4] In the Punjab, the Mochi, who is a Chamar, works in tanned leather.[5] He also grains leather. In some places the name Mochi denotes a *Mussulman* Chamar. Sometimes he is a weaver. In the west of the Punjab he is a tanner and leather-worker. In Ludhiana he is a weaver, and the name is almost synonymous with Julaha, but he does not intermarry with the latter. In the east of the Punjab, the Hindu Mochi makes boxes, saddles, and other articles of leather but not shoes. Some Punjab Mochis claim Rajput origin. The Bhanger of Kapurthala, a weaver, is an offshoot from the Mochi, but he does not intermarry with him.

[1] Sir Athelstane Baines, *Ethnography*, in *Gründries der indo-arischen Philologie und Atlertums Kunde*, 1912, p. 80.

[2] Nesfield, *A Brief Review of the Caste System of the North-West Provinces and Oudh*, p. 22.

[3] Crooke, *Tribes and Castes of the North-Western Provinces and Oudh*, Vol. III. p. 498.

[4] *Census Report, United Provinces*, 1911, p. 356.

[5] Rose, *A Glossary of Tribes and Castes of the Punjab and North-West Frontier Province*, Vol. III. pp. 123 ff.

CHAPTER II

SOCIAL AND ECONOMIC LIFE

As a rule the Chamar chooses his wife locally, outside his own village group, but within his own sub-caste. Although the sub-castes are essentially endogamous groups, marriages are occasionally arranged between members of different sub-castes. For example, Dhusiyas and Kanaujiyas intermarry,[1] and Jatiyas and Kaiyans sometimes do.

Again, the restrictions between endogamous groups may apply only to the giving, not to the taking of wives. Thus, Kurils will take Dohar girls in marriage, but will not give their daughters to Dohars. In such instances the Kuril settles with the *birādarī* by giving a feast; and, indeed, nearly all infringements of marriage regulations are usually adjusted by the panchayat's ordering the payment of a fine or the giving of a feast.

Occupation may become a bar to marriage, sometimes even within the endogamous group. Thus, those who remove manure and night-soil cannot intermarry with those who serve as grooms. Rai Dasis (in the Punjab) will not marry with Jatiyas who skin dead animals. Jatiyas in the Delhi territory, who work in the skins of "unclean" animals, are refused marriage by some clans of the Sutlej.[2] In some places Kurils who tan do not marry with Kurils who make shoes.

Within the sub-caste there are smaller exogamous or "family" groups (*got*, *kul*) which bear the name of some mythical saint, hero, or other person; the name of some

[1] Crooke, *Tribes and Castes of the North-Western Provinces and Oudh*, Vol. II. p. 194.

[2] Ibbetson, *Census Report*, *Punjab*, 1881, p. 181.

village or locality; or a name having reference to some totem. Marriage between members of the same exogamous group is prohibited. The *chacherā-mamerā-phuphерā-mauserā* law, which prevents a man marrying anyone in the line of his uncle or aunt on either the male or the female side,[1] is somewhat loosely observed; but the practice usually followed is that, so long as any relationship, however remote, is found on either side, marriage is forbidden. In some places a marriage is not arranged with any family from which a mother, a grandmother or a great-grandmother has come.[2] A man may marry two sisters, but in general may not have them both as wives at the same time, and the second sister must be younger than the first. He may not marry the daughter of a brother-in-law. Marriages are always arranged by the parents or relatives of the parties, and women are never contracting parties. Of course, the female relatives have a voice in the discussion of the marriage arrangements, and their opinion carries weight. Marriage is considered a sacrament and not a contract. Still, in some places, a bride-price as high as twenty or thirty rupees, and occasionally as high as one hundred, is paid; but the amount exacted is usually that fixed by custom. Nowadays this price generally takes the form of a contribution made by the groom's family towards the expenses of the wedding. Besides money, it includes gifts of clothes, food, sugar (*guṛ*), cooking utensils, and ornaments. A marriage is binding when the ceremony is performed, even if the consent of the parties has not been expressed or implied; but the consent of the relatives of both parties to a marriage must in every case be obtained. In case the marriage does not take place until after puberty, or where some other unfortunate circumstances have occurred, the bride may be given away. Daughters are married in order of seniority. When a girl may not be married, on account of some infirmity, or for some other equally valid reason,

[1] *Census Report, United Provinces*, 1911, p. 212.

[2] Crooke, *Tribes and Castes of the North-Western Provinces and Oudh*, Vol. II. p. 174.

the younger sister is allowed to marry. The younger sister may be married first, if the older is already betrothed.[1]

Under the principles of concubinage and polygamy, the practice of keeping more than one woman is common. There is no general objection to polygamy, provided a man is financially able to support more than one wife. Where the first wife is barren, a second marriage is usually sanctioned by the council. Furthermore a man may buy a widow or a younger woman. Widow-marriage often contributes to polygamy, especially where the younger brother takes the widow of his deceased brother. Although a second wife is often bought, she is not always regularly married. In some places, when a man takes a second wife, the first leaves him, and desertion under such circumstances is recognized as according to tribal custom. If the second woman live with the man for twelve years, she will have the same rights as the first. If the husband die, and the two women live at peace, both will inherit, provided he make a will. Rival wives, however, as a usual thing, do not get on together, and the quarrelling arising out of this condition has a special name, *sautiyā dāh.* There is a saying, "Even a co-wife of wood is an evil."[2]

Concubinage (*lauṁḍī, baṁbdī, rakhnī, rāṁdī, biṭhāī*) is widely practised, especially where men are able to support a large establishment, and the practice is not considered wrong. Two or three concubines are quite common, and some keep even more. They are obtained by purchase.

Among the Chamars early marriage is all but universal. The betrothal is very early, often in infancy, and marriage is usually as early as the eighth year. Any time between the weaning of the child and the eleventh year is considered proper for marriage. However, the age for the consummation of marriage is pretty generally recognized as that of puberty. Under special conditions, when the

[1] *Census Report, Punjab,* 1911, p. 268.

[2] Crooke, *Tribes and Castes of the North-Western Provinces and Oudh,* Vol. II. p. 175.

bride is an orphan, or when her parents are in financial straits, she may go to her husband's home at an earlier age. Usually the marriage is consummated when the groom is from sixteen to eighteen years of age and the bride from twelve to fourteen. The last Census returns for the United Provinces show that ninety-eight per cent. of all Chamar girls over fifteen years of age are married. The general practice of the caste may be gathered from the description of the marriage ceremonies. In 1891 Chamars were included in the group in which infant marriage most widely prevailed.

There are special forms of marriage contracts which may be mentioned here. One is marriage by exchange (*waṭṭa satta, gurāwat, adla badla*), where each family gives a girl in marriage to a son in the other. This is done to save marriage expenses, and is practised amongst the poor. Another[1] form of marriage is that in which, like Jacob, a boy serves a certain number of years for a wife. This is called *ghar jawāī*, and is sometimes arranged when a man has no son. The marriage relation may exist during this time.

There are in the Chamar marriage-ceremony many interesting survivals of marriage by capture. Among these are the bridegroom's coming mounted on a horse, if he can afford it, or in an *ekka*, or in a *ḍoli*, or a *bahlī*; in his carrying a sword, or something to represent it; in the *barāt* being composed of men, and in their stopping outside the bride's village; in the mock fight between the two parties at the bride's door; in the bride's being carried away in some sort of equipage; in the pulling down of one of the poles of the marriage pavilion or the shaking of it by the groom's father, and in the shaking of it by the groom; in the weeping of the bride; in the show of violence on the part of the bridegroom; in the mark of the bloody hand at both houses; in the fact that at the *pherā* the bride wears nothing belonging to herself, but things given by the groom's relatives; in the hiding of the bride; in the bringing of false brides and in other jokes at the

[1] *Census Report, Punjab*, 1911, p. 286.

expense of the groom and his party; in the fact that the bride's mother makes a mark in red on the groom's father's shoulder; in that the boy's village is tabu so far as drinking water by the bride's father and elder brother is concerned; in the fact that all the words denoting male relations by marriage are used as terms of abuse (*e.g.*, *susrā*, *sālā*, *bahnoī*, *jawāī*); and in the use of abuse directed by the bride's women-folk against the groom's relatives and friends all through the wedding ceremonies.

Chamars in 1891 were included in the group in which widows were comparatively few. The following table, taken from the census of 1911,[1] shows that the marriage of widows between the ages of twenty and forty is almost universal.

	All Ages			0—5			5—12			12—20			20—40		
	Un-married	Married	Widow	Un-married	Married	Widow	Un-married	Married	Widow	Un-married	Married	Widow	Un-married	Married	Widow
Males	408	522	70	993	7	..	896	100	4	478	492	30	89	836	75
Females	302	547	151	988	11	1	783	211	6	130	839	31	15	885	100

This remarriage of widows is legal and the tribal council may declare the children rightful heirs. The limits for such marriages are the same as for virgins. If the widow be young, and there be a younger brother of her former husband, of suitable age, they usually marry. There are traces of the levirate, in the right of the younger brother to take the widow in marriage. There is no idea of raising up seed for the dead brother. If the widow have brothers-in-law (brothers of her late husband), she must marry one of them, unless they choose to sell her, or make another arrangement for her. An older brother may take her. She may be married to the husband of an elder sister provided the latter be willing, or if the latter has died. If she is old enough to decide for herself, and if she has a child, her consent to the arrangements is taken;

[1] *Census Report, United Provinces*, 1911, pp. 244 and 245, shows number per 1,000 of each sex.

otherwise her relatives will decide. No ceremony is performed. The children by the former marriage may remain in the father's family, except in the case of an infant. Sometimes the woman takes all of the children with her, but then they do not inherit from their father. The settlement of the inheritance is usually made by the council. If there be no younger brother of suitable age, she may marry someone, usually a widower of the tribe, by an informal rite, but not by the *śādi* ceremony.[1] If she marries outside of the family, the bride-price must be paid to her former husband's relatives, and she loses the property and the children by the previous marriage. If the groom is not a widower, some form of mock marriage may be performed. By this ceremony the groom and the bride are placed upon the same level. It seems as if the widow were inherited by the *levir*, or bought by the outsider; as if she were property to be inherited, or to be sold. Of course her marriage is arranged for her by her own family, and the family of her late husband must agree to the marriage.

As a caste rises, the remarriage of widows and the levirate disappear together. For example, well-to-do Chamars in Cawnpore are prohibiting widow-marriage. Young widows (children) are mere household drudges, and are often ill-treated, poorly fed, and generally neglected.

Divorce is common. A man with the consent of the panchayat may turn his wife out for unfaithfulness, but she cannot get a separation on the same ground, if he feed and clothe her properly. A woman may desert her husband if he take a second wife. Impotency proved to the satisfaction of the council is another valid reason for a wife's abandoning her husband. In some places, a woman may not secure a divorce on the ground of disease or physical defect in her husband, provided his relatives continue to support her. The discovery of physical defects in the bride after marriage would be sufficient grounds for a divorce; and if a separation occurs on such grounds, the husband is usually satisfied if the

[1] This agrees with Manu.

marriage fee is returned. The divorced parties may marry others. Separation for adultery if the woman does not stay at home, and also for certain forms of disease, such as insanity, may be sanctioned. As a usual thing, the woman who is thus turned out of doors by her husband is either abandoned or sold. If she be sold, she may be married by the *sagāī* rite, and the issue of such a marriage can inherit. The principal causes of separation are when the woman leaves her husband and returns to her parents and when she goes to live with another man. In both cases the former husband receives back his wedding expenses. Divorce is legalized by the panchayat. Sometimes the woman breaks a straw as a sign that her marriage has been dissolved.

Traces of the matriarchate are seen in the following facts: The marriage is arranged by the mother's brother, or the mother's sister's husband, or these relatives play an important part in the negotiations; the father's sister's husband, *negī*, has duties at the wedding; there are other similar relationships involved. Again, the uncle's (mother's brother's) consent to the marriage is necessary, and he sometimes receives all or part of the bride-price. In other places his privileges are confined to the making of certain gifts, such as earrings, the wedding clothes for all the family, and a certain number of rupees towards the wedding expenses, and the furnishing of the dinner for the barat. These privileges are not always obligatory. There are other duties in connection with the funeral rites and the practices connected with the birth and early years of children, which point in the same direction.

Social intercourse is lax and moral standards are exceedingly low. Irregular unions, such as concubinage, both inter-tribal and extra-tribal, are admitted by the Chamars. Where sentiment is against such practices the payment of a fine removes disabilities. Sexual irregularities are common. When they are brought to the notice of the council, they are punished by fine. A man may leave his wife and take another, yet through the panchayat he may demand his former wife back again. If a woman is discovered in adultery, a fine and a feast are required by the panchayat or she is out-casted. In case a widow becomes pregnant, abortion is resorted to, or some marriage is arranged, or she may be sold. If she names the father of her child, or if the panchayat discovers him, they are required to marry, but both are ostracized for about a year, after which the panchayat may recognize them and their union. If this irregularity be with a man of another caste she is excommunicated. Pre-marital immorality is common, and, if within the caste, is much less serious than if detected with outsiders. A pregnant girl simply names before the panchayat the man concerned, and he must take her as his wife, that is, if they are not of the same got, and he is unmarried; otherwise he must pay a fine. This always includes cash and a feast. She will then remain with her parents, or they may arrange a wedding for her or turn her out. These matters are sometimes severely dealt with. The children of such irregular unions have no property rights. Again, the guilty man in such a case may pay a bride-price and she may marry someone else. The sale of a woman is common when she gives trouble, or is unhappy, or lazy, or disobedient, or if she be a bad character. The purchaser takes her by means of the less formal marriage ceremony. In the Punjab Chamar women are sold to Jāṭs, to Gujars, to some Rajputs, and to Mohammadans as wives.[1] Some bring as high a price as 200 or 300 rupees. These are usually women of the poor. Women are sometimes gambled away. In case of children born from irregular marriages, if the woman be of a higher caste or rank than the husband, the children have full caste rights, but restricted inheritance or no inheritance at all. In other cases, the offspring belong to the caste, or tribe, of the father, except when the mother is a Mohammadan, or of a lower caste. There are certain kinds of laxity that are common. A visitor occasionally has liberties with the host's wife or daughter. But this is not considered "good." The relatives of the

[1] *Report, Census of India*, 1911, p. 378; some of these are from the United Provinces.

husband take certain liberties (*haṁsnā-khelnā*) which usually do not extend to immoral acts. There is sometimes prostitution in the home, and sometimes the wife is hired out. Women sometimes exchange husbands secretly. A woman may go and live openly with another man and still be received back. Sometimes, when men are in the relationship of very close friends, having vowed friendship on rice from the temple of Jagannāth, they will each place his wife at the disposal of the other.[1] In the Central Provinces Chamar women are hired for the *śakti mārg* ceremonies, and women of the Mādigās and Chakalliyans of the South are chosen for similar rites. During the year at certain festivals, such as the *Holī*, the *Dewālī*, and the *Sāwan*, there is great sexual license. Not only are the songs of these festivals obscene beyond imagination, but the people give themselves up to unlimited excess.

There are other social customs, more or less objected to, but often allowed and not considered wrong, which are gradually disappearing under modern conditions: such are the *jus primae noctis* of landlords and *gurus*. The zamindar often has liberties with the Chamar's wife in consideration for his payments to the Chamar. The sais's wife often gives immoral services where her husband is employed in the towns or cities. Furthermore there are certain customs within the caste which are most debasing. "Formerly, when a Satnami Chamar was married, a ceremony called *Satlok* took place within three years of the wedding, or after the birth of the first son, which Mr. Durga Prasad Pande describes as follows: It was considered to be the initiatory rite of a Satnami, so that prior to its performance he and his wife were not proper members of the sect. When the occasion was considered ripe, a committee of men in the village would propose the holding of the ceremony to the bridegroom; the elderly members of his family would also exert their influence upon him, because it was believed that if they died prior to its performance their disembodied spirits would continue a comfortless

[1] Russell, *The Tribes and Castes of the Central Provinces of India*, Vol. II, p. 412.

existence about the scene of their mortal habitation, but if afterwards that they would go straight to heaven. When the rite was to be held a feast was given, the villagers sitting round a lighted lamp placed on a water-pot in the centre of the sacred *chauk* or square made of lines of wheat flour; and from evening until midnight they would sing and dance. In the meantime the newly-married wife would be lying alone in a room in the house. At midnight her husband went into her and asked her whom he should revere as his guru or preceptor. She named a man, and the husband went out and bowed to him, and he then went in to the woman and lay with her. The process would be repeated, the woman naming different men until she was exhausted. Sometimes if the head priest of the sect was present, he would nominate the favoured men who were known as gurus. Next morning the married couple were seated together in the courtyard, and the head priest or his representative tied a *kanthi* or necklace of wooden beads round their necks, repeating an initiatory text. . . . . It is also said that during his annual progresses it was the custom for the chief priest to be allowed access to any of the wives of the Satnamis whom he might select, and that this was considered rather an honour than otherwise by the husband. But the Satnamis have now become ashamed of such practices, and, except in a few isolated localities, they have been abandoned."[1] The practice has not been entirely abandoned.

The probability is that female infanticide is not practised by the Chamars, although female infants are neglected, often deliberately. When food is scarce they suffer most. But other reasons will account for the disparity in numbers between males and females. The woman is more subject to plague and malaria, owing to her domestic duties and to her closer confinement in the house. Besides this, unsanitary and unclean methods of midwifery are the cause of a good deal of female mortality. Furthermore, the practice of infant marriage reduces the vitality of women and subjects them to many dangers. Yet, when

[1] Russell, *The Tribes and Castes of the Central Provinces of India*, Vol. I. pp. 311, 312.

all the disabilities of women are taken into account, the proportion of females to males is high. In the United Provinces, there are, among the Chamars, 958 females to every 1,000 males.[1] This is above the average for the Provinces for the whole population.[2] For Bihar and Orissa the proportion is 1,153 females to 1,000 males, for the Central Provinces and Berar 1,035 to 1,000, and for the Punjab 846 to 1,000.

Not only is the moral standard of the Chamar low in respect to social purity, but also in matters of excessive use of narcotic drugs and intoxicating beverages. Drunkenness is a caste-failing and forms a prominent element in many domestic and religious customs.

The Chamar is not fastidious about his food. He eats the leavings from nearly all castes, except the *Dhobī* and Dom. The death of a buffalo or of a cow in the village is his opportunity for a feast. This is almost universally true, although there are sub-castes some of whose members do not eat carrion, and the number of such is growing. There is, however, not a single sub-caste that is free from this practice. Sometimes the chief men of a sub-caste may refuse to share in such food. Furthermore, many Chamars eat pork. In general the flesh of fowls and of cloven-footed animals goes to the Chamar,[3] while that of such animals as do not divide the hoof goes to the Dom or *Bhaṁgī*. The Chamar in general will not touch the carcasses of ponies, camels, cats, dogs, squirrels, and monkeys. Those are delegated to the Bhaṁgī. Strange as it may seem, in some places (*e.g.*, the Punjab, in Hindu communities), while he eats dead cattle, the Chamar may be excommunicated for eating beef.[4] In Mohammadan communities there is no such scruple.

His ordinary food consists of bread made from the flour of the cheaper grains such as gram, barley, and millet, and of such grains as he may get as pay for labour at harvest-time. His regular meal is at night. He has some grain

[1] *Census Report*, *United Provinces*, 1911, p. 204.
[2] Which is 915/1,000.
[3] Ibbetson, *Census Report of the Punjab*, 1881, p. 320.
[4] *Census Report*, *Punjab*, 1911, p. 111.

in the morning and *sattū*[1] at noon. "He considers that his full ration would be two and a half *pakā* seers of grain or about three and a half Government *sīrs*. Some days he gets only one seer and sometimes one and a half seers. A large part of his diet consists of whatever vegetables, such as leaves of gram, mustard, etc., his wife and children can pick up in the fields. His rule is to mix from two to four *chhatāṁks* of flour in about two and a half seers of vegetables. These are all boiled down into a mess and eaten hot with the balance of the flour made into bread."[2]

Some groups, as for example the Jaiswars, refuse to eat any food prepared by others. It is difficult to say just how far these distinctions are observed, but in general the main sub-castes do not eat or drink or smoke together. Chamars will accept cooked food from members of their own sub-caste and from those sub-castes which are of a slightly higher social status. For example, a Chāmar will accept food from a Jatiya, but the reverse is impossible. There is a gulf between these sub-castes, not only determined by occupation, but by other considerations as well, for a Jatiya plasters the place where he cooks his food with cow-dung, while the Chāmar does not. The former will eat goat's flesh but not beef, while the latter has no such scruples.

The rules pertaining to the drinking of water are similar to those with reference to eating. For example, a Jatiya, while he will not drink water in the house of a Chāmar, will take the latter's *loṭā*, clean it, draw water with it from the well of the Chāmar and drink it. The vessel in which the water is brought must belong to a member of the caste. Women draw and carry the water required for household purposes. A Chamar will accept spirituous liquors from the hand of a higher sub-caste man but not from that of a lower. If he drink from the hands of a member of another sub-caste, he will require a separate cup; but if those who drink together are of the

[1] Flour made from parched grains, such as barley and gram.
[2] See Morrison, *Industrial Organization of an Indian Province*, p. 197.

same sub-caste, they will drink from the same cup. The rules governing smoking are quite similar. Members of the same clan will smoke together, but if men of different sub-castes are present, each group will have its own *huqqā*. Other castes do not smoke with them. Chamars will smoke together, using the same *chilam*.

The men and women of the home do not eat together. The women prepare the meals and eat after the men have finished. Only in times of sickness do the men condescend to do much household work.

Since the caste is largely shut up within its own limits, social intercourse is almost wholly a caste matter. Higher castes do not mingle with them and the Chamars will not associate with castes of lower social status. They observe caste rules governing marriage and commensality, and are said to conform to Hindu practices rather more strictly than better-class Hindus.[1]

Chamars will not accept food from Mohammadans. When, however, they are out-casted, they will eat anything.

In some places sections of the caste are slowly securing a higher social position by adopting the usual methods employed in India.[2] Those who are well-to-do, are making an effort to seclude their women, are prohibiting widow-marriage and are discouraging the more disgusting and heterodox practices of eating pork, beef, carrion, and the leavings of food of other castes. Such sections are slowly separating themselves from the main caste and from the name "Chamar." But, as a whole, the caste still occupies a position on the very outskirts of Hindu society.

The Chamar has a well-organized and influential council, or panchayat. It is greatly feared, and exercises a very strong influence over its constituency. In its simplest form it consists of the whole village or *mahallā* group, is conterminous with the sub-caste to which the Chamar belongs, and consists of all the men under its jurisdiction. In its less extensive form it is a body in which

[1] *Census Report, United Provinces*, 1911, p. 123.
[2] See *Ibid.*, 1911, p. 119.

the families of a village group are represented, or it may be composed of all the old men. Usually the representation is by families. There may be a sub-committee, often composed of five persons, which guides and rules the larger body. Amongst the Chamars, as amongst most of the functional castes, the panchayat is a permanent body, that is, the headman (*chaudharī, sarpañch, pradhān, methā, sardār, mukhiyā, mānjan*) is elected for life. The office is usually hereditary. When a chaudhari dies leaving a minor son, his relatives usually act for him during his minority, allowing him to announce decisions. When it becomes necessary, someone else may be chosen to succeed the father; but this would probably be some other member of his family. In Rajputana there are places where the raja appoints a chaudhari. Continuation in office depends upon good conduct and competency. A vice-president (*nāib-pañch, ḍārogā*), or summoner of the council, is a more or less permanent officer, chosen by the panchayat. He is sometimes called the *chhaṛidār*, or mace-bearer. He serves as an assistant to the headman. For his services he gets a small money fee, sometimes about half what the chaudhari receives. There is a chaudhari in every community or village, and, oftentimes, a sarpanch or chaudhari, who governs a group of villages.

The investiture (*pagṛī ḍālnā, pagṛī lagānā*) of the chaudhari with his pagri (turban) is a serious matter, for it is his official inauguration into an office which is of great importance in the social and economic life of the Chamār. The whole village group performs this act as a sign that they have chosen him and have entrusted him with their interests. If the same man be chosen for two or more villages, or mahallas, each will give him a pagṛī. Before the investiture a careful examination is made as to the candidate's fitness for the office and as to his character. If the Chamars are satisfied on these points, a day is fixed for the ceremony. At the appointed time the whole group assembles for the purpose. First, with the use of a lota and a basin, there is a general foot-washing ceremony. This is followed by a fire sacrifice (*hom*), after which the candidate is conducted to a conspicuous place in the midst of the assembly. A white pagri together with one and a quarter rupees and a cocoanut are then presented to him. Occasionally a *ṭīkā* is made on his forehead with *haldī*. Sometimes one, or five, rupees are placed in the pagri. Then the assembly greets him as chaudhari. A great feast, in which both rice and sugar are included, follows. There is an idolatrous phase to this dinner similar to that observed in the death feast. There is an excessive use of country spirits. Women do not take part in these festivities. The expenses of the feast are met by a public collection. The candidate himself gives a preliminary feast to the group. His official perquisites are certain fees and a percentage of all fines connected with trials and a share in the feasts. His office brings him in a considerable income.

All ordinary matters are brought before the local body. But, when cases of major importance are to be considered, several panchayats may be called together; that is, the headmen of several villages, each with a number of influential Chamars, meet with the *pañch*, in the village where the case has been brought. In very grave matters representative men from widely scattered areas may be called together. Each sub-caste has its own independent council, and, with rare exceptions, different sub-castes do not meet in council. However, one or more influential men (panch) of another sub-caste may be called in for advice. Cases are known, as when the interests of the whole caste are involved, of a general meeting of representatives of all the chief local sub-divisions of the caste. Such a council is called "*sabhā*" and is quite modern. Such a one was held in Bijnor some years ago. In some places in the Punjab, and in the United Provinces also, there are village panchayats in which the Chamars are represented.

The jurisdiction of the panchayat is local, but other panchayats may enforce its findings. The panchayat exercises jurisdiction over the following classes of cases: (1) Of illicit sexual relations, such as the discovery of a pregnant widow, of adultery, or of fornication. If the matter is not well known, the parties are let off with a fine and a threat, but if the irregularity be a public scandal, a trial must be held. (2) Of the violation of the tribal rules concerning commensality. (3) Of matrimonial disputes, such as the sale of widows and cases where a girl is not given in marriage after the betrothal. (4) Of petty quarrels that would not come under the cognizance of the Government Courts, such as false witnessing, fighting and quarrelling. (5) Of disputes about small money transactions and debts. (6) Of cases connected with hereditary rights; and (7) of matters affecting the welfare of the caste.

There are certain occasions, such as caste dinners of all kinds, when persons take advantage of the gatherings to bring matters before the panchayat. Council meetings are avoided at marriages, but are often held during funeral services.

Meetings of the panchayat may be summoned by either party to a dispute. Cases are usually brought before the whole village group by the offender who wishes to clear himself. But the headman or some other party may lodge a complaint. The person who calls the council must furnish tobacco enough for the whole company and a huqqa. He must also pay a fee of one and a quarter rupees to the chairman, who will not take up the case unless it is paid. This fee is usually spent for spirits. The village group is called together and the case involved is thoroughly talked over. All evidence is oral. Anyone may speak. Often an oath is taken over Ganges water, or upon the plough, or with a son in the lap. This is resorted to in cases when it is difficult to reach a decision or to get at the facts. After a full discussion five men are chosen to give a decision. There is no custom which necessitates the choosing of the same five men in case after case. The decision, which is pronounced by the headman, is binding. Decrees are not published, except in special cases. When the council finds a person guilty of the offence charged, it imposes a penalty which usually takes the form of a fine. This may be levied in rupees, or may be an order for the offender to entertain the clansmen. The fine may be any reasonable amount, but the sum collected seldom exceeds five rupees. The fines in certain classes of cases are fixed by custom. Until the fine is paid, or the feast given, the offender is not allowed to eat or drink with his clansmen. Another and a more serious result of conviction is that until the ban is removed all marriage alliances with the family of the offender are barred; and, if anyone marries a member of such a family, he at once becomes liable to the same punishment as that which they are undergoing. It is very seldom that the process of excommunication has to be used to enforce payments. The fines are spent in the purchase of spirits for the members of the tribe and in feasting them, or for some such purposes as the digging of a well. A certain proportion, however, of the fines collected is the perquisite of the chaudhari. Besides this a certain percentage of the fines is often set aside as a sinking fund for special purposes, such as the hiring of lawyers when trials occur in the Government courts. Some unusual punishments include the sending of persons on pilgrimage, requiring them to solicit alms, and various forms of degradation. Sometimes a beating with a shoe is pronounced as a punishment; and again the shoes of the whole party are placed upon the head of the offender. For discovery in sexual irregularities the parties are sometimes taken to the bank of a tank, or river, where their heads are shaved in the presence of the panchayat. They are then made to bathe. The shoes of all the company are then made into two bundles and placed upon the heads of the guilty pair and they are made to promise not to repeat the offence. Frequently the convicted party is bound to a tree and beaten. If a Chamar entice away the wife of a clansman, in addition to the punishment inflicted by council he is obliged to pay her marriage expenses. Even excommunication resulting from irregular marriages, and the punishment of the most grievous offences may be remitted by the payment of a fine. Becoming a Christian does not necessarily result in excommunication. Although he will, as a Christian, abjure caste practices, he is not excluded from social intercourse with the sub-caste from which he came. But a Chamar who has turned

Mohammadan is permanently excluded from his clan. In some places, where the Christian is considered by the caste as a social outcaste, he may be reinstated by the payment of a fine. The amount imposed will depend upon the financial ability of the outcasted party. Where a whole village which has become Christian desires to be reinstated in the biradari, an amount, determined by the financial resources of the village, is paid through the chaudhari to the head chaudhari of that particular part of the country. There is no ceremony of re-instatement; they simply resume the exercise of privileges amongst which *huqqā-pānī* and *śādi-biyāh* are the most esteemed.

In some cases heavy penalities are imposed. For example, a chaudhari was outcasted for twelve years for showing partiality to his brother (the punishment was afterwards reduced to a fine by a council of panchayats). Another Chamar, who disgraced his caste by begging, was outcasted. His son was reinstated by paying a fine of four rupees and feasting five Brahmans.[1] Some others, who were in a court convicted of poisoning cattle, were excommunicated for twelve years. They offered 500 rupees to be reinstated but in vain.[2] In another case two Chamars were fined ten and six rupees respectively for removing dead animals from the house of another Chamar's clients; and the husband of a Chamar woman who worked as midwife for another Chamar's client was fined five rupees.[3]

The work of the panchayat is of great importance. It relieves the courts of a great many petty cases, on the one hand; and, on the other, it is of great regulative value in the life of the village group.

There are certain hereditary rights which are the privilege of a certain Chamar family (or families) in each village.[4] These rights, called jajman or *gaukamā*, are

[1] *Census Report, United Provinces*, 1911, p. 340.
[2] *Ibid.*, 1911, p. 341.
[3] *Ibid.*, 1911, p. 342.
[4] See Morrison, *Industrial Organization of an Indian Province*, pp. 179, 180, 194-197, from which a good deal of this discussion is taken. See also Crooke's *Tribes and Castes of the North-Western Provinces and Oudh*, Vol. II, p. 175.

carefully guarded. In return for these perquisites the Chamar gives regular services to the landlords. The circle of clients from whom he receives these privileges expect him to remove dead cattle, to prepare leather from the hide, and to furnish a certain limited supply of shoes and other leather articles. Besides the dead cattle, which belong to him by right, he gets a fee of from ten to twelve seers of grain for curing the hides of the animals that die. From the hide he sells one pair of shoes to the zamindar for two and a half seers of grain. The rest of the hide is his. Occasionally he is expected to mend, or even to make, shoes for nothing. In some places he can claim the hide without the obligation of furnishing anything. These rights in respect to hides are now being questioned and in some cases denied altogether; but the landlord is obliged to make some concession, which is usually in the form of privileges of cultivation. Besides the rights connected with leather, the Chamar receives certain small privileges, such as fuel and grass from the village lands and gifts at stated festivals and on other social occasions. He is expected to work for his clients upon demand, but receives certain definite gifts of grain at harvest-time.

The Chamar's wife has her clientèle, as well, for whom she acts as midwife, and for whom she performs various menial services at marriages and festivals, such as collecting wood, bringing earthen vessels from the bazar, supplying cow-dung and grinding grain.

The following summary of the Chamar's perquisites as a labourer in rural districts is substantially from Morrison. When grain is threshed, the Chamar gets twenty seers at each harvest per plough in consideration for repairing the well-water bags, for providing leather straps and whips, and for helping to clean the grain. The light grain and sweepings of the threshing-floor are his perquisite in consideration for the help that he gives in threshing and in winnowing. For work in irrigation his wages are often one and a half annas per day. He receives three bundles of the cut crops each day during the harvest. These are large or small according to the amount of work that he does. As a ploughman his wages are a daily portion of grain from one and a half to two seers of *rabi'* grain, or pulse, at mid-day, which represents half a seer of sattu, and for fifteen days during seed-time he will get an additional allowance of one seer a day. The practice of paying the Chamar in kind is being discontinued in certain parts of the country. This is due to changing economic conditions. In former days he used to sell his grain in the markets and purchase the things which he needed for himself. Women and children do the weeding, for which each gets a seer of grain, or such an amount as is fixed by custom, and, sometimes, an extra allowance. At reaping-time all hands receive one good bundle for each sixteen small bundles gathered for the landlord. At earth-work an able-bodied man earns two seers of grain and half a seer of sattu and an additional handful of grain to start with in the morning. For carrying flags and doing other services in wedding processions, both father and son receive gifts after the wedding and an allowance of food during the festivities. The Chamar often gets the old clothes and blankets which the zamindar wishes to give away. These fees and allowances are scarcely more than illustrative. The actual amounts vary, and the whole system of perquisites is in a somewhat unsettled condition.

For services as midwife, the Chamari receives food and presents. These will be more or less according as the child happens to be a boy or a girl, or the firstborn. Her usual perquisite is a new *sāṛī* and four annas in cash. But these fees have been considerably increased in recent years. It is to be noted, however, that there are areas where this work is done by other and lower castes; and further that, in the same sub-caste, in some areas the women are engaged in this profession, while in others they are not. The practice of midwifery is looked upon as most degrading. The women who follow this profession employ methods of the crudest sort. Sanitary conditions are almost entirely neglected, and no attempt is made to prevent infection. A considerable percentage of the mortality amongst women is traceable to the work of the midwife. The ceremonies of the sixth day are to a certain extent directed against tetanus, which is prevalent especially amongst babies. The conditions under which the mother is confined are most unfavourable. The room is kept close, and she and all things within the room are considered unclean. A fire is kept burning constantly, and very often the atmosphere is laden with the heavy smoke of incense. Such things as red peppers and old leather are amongst the articles that are cast into the fire. The whole technique of the practice of midwifery is directed by custom and superstition; and the evil smells and the other barbarous practices connected with the lying-in room are designed to beat off demons of disease and of destruction. Unfavourable signs, such as fever, are the occasion for the practice of magic and the burning of such things as give off most distressing and oppressive odours.[1]

There is a real sense in which the Chamar has to do work for which he receives no compensation. These conditions are well known and need no proof. A characteristic illustration is found in the following incident. A young Chamar left his section of the country and took up service. He became fairly prosperous and felt that he had risen in the world. He concluded to pay a visit to his native village. There he chanced upon his old master, who said, "Give me that umbrella. You have no use for it. I will give you eleven annas." So, taking it, the landlord said, "Go to work with the plough to-morrow." The next morning the landlord's servant appeared and forced the Chamar to go to work. In the evening the young man received three pice for his day's work. He realized then that he was only a Chamar after all. As a class, they are oppressed and they live in continual fear, especially of the zamindars, and far from having the comfortable environment pictured in *Industrial Organization of an Indian Province*, their lot is a hard one. They are constantly harassed by demands of all kinds. Men are needed for some odd job and a request is sent to some officer. A peon goes to the Chamar section of the village or town, and impresses the number of persons

[1] For further details see Chapter III.

required. They are supposed to receive wages for their services, but they are more or less at the call of others, no matter what their own interests may be. There are certain duties which they must perform for Government and for the landlord, and for these they receive certain privileges related to the land. There are, however, many instances where they are required to work without pay, under the direction of petty officers.

Tanners are more common in the Meerut, Agra, Rohilkhand, Allahabad, and Lucknow divisions, and less common in the Benares, Gorakhpore, and Fyzabad divisions. Furriers are found only in Saharanpur and Bara Banki.

A catalogue of the different kinds of work which the Chamar performs, shows that he belongs to the great class of unskilled labour. He is a grass-cutter, coolie, wood- and bundle- carrier, drudge, doer of odd jobs, maker and repairer of thatch and of mud walls, field-labourer, groom, house-servant, peon, brickmaker, and even village watchman. He is the common labourer along the railways and in the great cities. He does a good deal of weaving. The contractors who undertake petty repairs in the towns and cities are often Chamars. He repairs the underground rooms and makes the bins where grain is stored, and prepares the threshing-floors. Besides, he beats drums, rings bells, and blows trumpets at weddings or when cholera or other epidemics are being exorcised from the village. He also makes musical instruments. Some sub-caste names are illustrations of occupational functions; for example, *Mochi* (shoemaker), *Chāmkatiyā* (leather-cutter), *Châmar* (leather-maker), *Chamār māṁgtā* or *Māṁgatiyā* (beggar), *Kātuā* (leather-cutter), *Tāṁtuā* (maker of leather thongs), *Zingār* (maker of saddles), and *Nālchhinā* (one who cuts the navel cord).

It is as a tanner and worker in leather that the Chamar obtains his name. Besides making the thongs, baskets and other articles used in husbandry, he is a maker and cobbler of shoes. He furnishes not only the shoes made according to country patterns, but also, and in rapidly increasing quantities, shoes and boots made on English models. He is also a dealer in hides. In the Central Provinces he has, in some instances, become a dealer in cattle.

But the Chamar is not now chiefly a tanner and a worker in leather. The census returns for the United Provinces in 1911 show less than 131,000 who reported their hereditary occupation as their principal means of livelihood, while but 38,205 reported leather-work of any sort as their subsidiary means of livelihood. But 26,112 actual workers who returned their traditional occupation as their principal means of livelihood, had some subsidiary occupation. 1,354,622 recorded their principal occupation as cultivation; 1,245,312 were returned as field-labourers, wood-cutters, etc.; 142,248 as artisans and workmen; 331,244 as labourers (unspecified); and 31,855 as domestic servants.[1] In the United Provinces the great majority of the Chamars are engaged in "the exploitation of the earth's surface." Similarly we find that, in the Punjab, they are an extensive class of low-caste cultivators; and that in the Central Provinces, the great bulk of the caste, namely, the Satnamis, do not touch leather at all. The figures from the United Provinces[2] show that only five per cent. of the Chamars are leather-workers; that seventy-eight per cent. of them exploit the earth's surface (*e.g.*, are cultivators, agriculturists, and labourers); that four per cent. are engaged in other industries; that two per cent. are occupied with transport and trade; and that nine per cent. are general labourers. In most occupations both men and women are engaged. Chamar women, besides performing the ordinary house duties, do an immense amount of work in the fields. This consists of weeding and other forms of lighter work connected with the care of the crops. They also do the husking and grinding and help in the winnowing. In addition to this they do a considerable amount of ordinary coolie work such as carrying produce to market, and the like. They do not, however, compete with the men, but rather supplement their work. In the hide industries the

[1] *Census Tables, United Provinces*, 1911, Table XVI, pp. 757 ff.
[2] *Census Report, United Provinces*, 1911, pp. 412, 413.

number of women-workers to one thousand men is one hundred and eighty-five.[1]

Economically the Chamar is a most valuable element in the population, and his function is the rough toil and drudgery of the community. Though nearly always a poor man, he, as a rural labourer, generally has plenty to do. His work is distributed over the year about as follows: For five months, June to November, he works in the field with a plough; for two months, November and December, he is engaged in reaping the *kharīf* (the autumn crops); during January and February he is occupied with *kachchā* buildings and other forms of earth-work; in March and April he is busy in gathering the rabi (spring harvest); and in May he does a little earth-work. Between times he does whatever work comes to hand. For the most part he is still in an almost hopeless state of degradation and serfdom. In large areas he is at the beck and call of others, and dares not lift his voice in protest lest he be beaten or driven from his village. However, economic changes are taking place, and Chamars are leaving the land to take up employment on the railways and in the industrial centres. In some parts of the country as many as twenty-five per cent. of them are away from home half the year. The result is an increasing demand for field-labour. Consequently wages have been enhanced. The recent increase in the value of farm products has resulted, in some instances, in the substitution of cash for grain as wages. This will eventually help the Chamar. The increased value of leather has led the landlords, in some parts of the country, to question the Chamars' traditional right to raw skins. But the landlord has been obliged to offer another form of compensation, and this has been in cultivating privileges. The Chamars' rights of occupancy are being obstructed in many places, and the laws which have been framed for his protection have not always secured him his just dues; still, the amount of land that is coming into his possession, both in the form of non-

[1] *Census Report, United Provinces*, 1911, p. 402.

occupancy and of occupancy rights, is slowly increasing. Some Chamars are owners of land, and in the Central Provinces, for instance, whole villages are possessed by them. Not only are they under the heel of the landlords, who they fear may deprive them of their cultivating rights and of their houses, but they are also under the influence of the *baniya* and the landlord, from whom they borrow to purchase seed-grain, leather, and oxen. Debt becomes a heavy shackle for them, and often the labour of their whole family is employed in satisfying the claims of creditors. As these people begin to discover their rights before the law, and as they gather courage, their position must improve. Not infrequently Chamars shift to other villages where conditions are more tolerable, or they appeal to someone who is willing to help them to obtain justice. These are encouraging signs. Still, the process which will lift him from dependency to independence is a long one, and as yet he has scarcely begun to move.

CHAPTER III

DOMESTIC CUSTOMS: BIRTH

BARRENNESS is looked upon as a great misfortune by Chamar women; and to remove this reproach they visit noted shrines and tombs and make offerings, including cocoanuts, *līchīs*, grains, and small sheets. Ashes taken from the smouldering log belonging to a holy man, and medicines obtained from *faqīrs*, are used as cures; and some women wear around their necks blue-black threads blessed by a *bhagat*, or wizard. Similar devices are employed in the effort to obtain a son. Under the direction of a wizard ants are fed daily with a mixture of sugar and flour; fish are fed with balls of flour; and the *pīpal* tree is watered daily for a year. Some vow to forego salt on Sunday, or for a given period. Women used to set fire to houses, believing that this would result in the obtaining of their desires. Seven or twelve houses had to be destroyed. The fear of imprisonment now acts as a successful check to this practice. Occasionally, a woman will secure by stealth, and swallow a piece of the umbilical cord of a recently-born male child, believing that she will thereby secure the mother's gift of fertility (of course the mother will become barren). Some women curse boys, hoping that they may die, for then there is the likelihood that the boys will be reborn as their own children. In desperate cases, when male (or even female) offspring is especially desired a bhagat is called in. He repeats spells and incantations over a cup of water, wags his head, and goes through various other antics, until he has obtained the desired "demoniacal" possession. He then places his hand upon the woman, gives her the water to drink, and promises her the fulfilment of her desires. The wizard receives gifts. Sometimes several bhagats are called in, and each performs his own magic.

Although Chamars believe in general that the knowledge of sex is one of the secrets of the Great Spirit, Brahmans are sometimes called in to prophesy as to the sex of the child. They use the chance methods of the fortune-teller. Some assert that there are signs which foretell sex. For example, if at the time of conception a man's right nostril twitches, the child will be a boy; if the left nostril, a girl. Again, if after conception, the mother goes to sleep upon her right side, a boy will be born; if on her left side, a girl. There are certain signs that indicate the sex of the child. If, in the later stages of pregnancy the right breast, or the right side of the mother, be the larger, or if she becomes thin, a son is sure to be born.

The desires of pregnancy, which they believe may begin immediately after conception, or from the fifth month, are thought to be the desires of the child, and must be granted, or the child will either die or fall under the spell of the evil eye. During pregnancy purgative and laxative foods are avoided; foods, such as oil, rice, and *uṛd*, which may cause, as they believe, abortion, are forbidden; and likewise foods, such as vinegar and spices, that might give trouble to the child.

Chamars are particularly exposed to the fear of witchcraft and of diabolical agencies generally, so they take every precaution to protect the prospective mother from evil influences. During the pregnancy the woman wears blue-coloured threads, given by a bhagat, around her neck, and a copper coin of the old mintage in her hair, and hangs charms, fastened with blue-black threads, about her neck and waist. She does not wear red clothes, but prefers white or black garments; she avoids blood; she keeps a knife under her pillow at night, and wears *hīṅg* (asafoetida) in her dress by day. She must not touch a woman who has had a miscarriage, and she must not have flowers taken into her room. A pregnant woman who is afraid that her child may die, will sell it to a neighbour for a trifle, or later she will give it a name that will serve to avert the evil eye and that will indicate that it is not worth the attention of demons.

If during pregnancy an eclipse occur, the woman must remain in the house, and she must do no work. If she does not remain perfectly quiet her child will be deformed. A circle of cow-dung is drawn on her abdomen. She must not be allowed to sleep. If she eat, her child will go mad; if she uses a needle, the child will be marked with a hole in the skin, usually about the ear; and if she uses a knife or scissors, there will be a cut upon the child's body, most likely he will have a hare-lip.

Not long before the time of parturition, and at other times as well, promises of offerings are made to various godlings and to the sainted dead, to insure a safe and easy delivery. For example, a vow is often made that, if the child is safely born, they will shave his head, and offer the hair to the Ganges. So, some time after the birth of a child, perhaps four or five days after purification, or during the sixth month, the child's head is shaved and the hair is wrapped in a *pūrī* (a thin cake of meal fried in *ghī* or oil), or placed between two puris, and cast into the Ganges. If the river is not near by, the hair may be buried in the compound, or, they may wait until some *mela* gives them occasion to visit the river. In this latter case, they will not only make the offering of the hair, but also they will offer the child to the Ganges, casting it into the river, leaving it unsupported for an instant, and then catching it up again before any harm comes to it.[1]. This may be repeated seven times. Occasionally the child is caught up by a Brahman and bought back by its parents.

When the birth-pains begin, the woman is given ghī to eat and water, in which urd has been soaked, to drink; or, a coin is washed in water and the liquid is given to the patient. A copper coin is placed in the woman's mouth, and pice are offered to the various godlings. At the right of the bed (*chārpāī*), upon which the woman will rest, barley is scattered for *Shastī*. At the door of the delivery-

[1] There may be in this act some reminiscences of an earlier barbarous practice.

room thorny branches of *bel* and of *nāgphanī* are hung to intercept evil spirits. A fire is kept burning constantly in the room near the door, and into it *ajwain* (seeds of a plant of the dil species) and other things are occasionally cast. Sometimes an old shoe is burned. If the birth-pains are excessive, or if delivery is delayed, men and women pound clods of earth together.

The woman sits on her heels on the ground during her accouchement and is supported by her female relatives. After birth, a song called the *sohar*, which is mostly an invocation of *Sītalā Māta*, is sung by the women of the neighbourhood. The singing is kept up more or less continually until the evening of the sixth day. The *tawā* (a sort of frying-pan) is beaten to protect the child from demons. In case a daughter has been born, the singing, or the beating of the tawa, may be neglected. The custom varies over the country, and in some parts almost as much protection is given to a girl as to a boy. Still, there is less rejoicing at her birth than at that of a son. In some places the mark of the hand in red-lead or in *gobar* (cow-dung) imprinted on the side walls of the house signifies that a son has been born. A line drawn on the wall all the way around the house signifies the same thing. Many devices used to protect the mother and the child are employed with greatest care if a son has been born. A net, or a branch of a *nīm* tree, or of the *siris*, and an iron ring may be fastened over the door. It is good to hang up a bunch of mango leaves over the door because it will attract some godling who will protect the child.[1] Charms are stuck on the walls of the house. A fire is lighted in the room near the threshold and kept burning night and day. As soon as the child is born, the mother's face is washed, and her forelocks, or her hair, are let down. Then the navel cord is cut, and the child is rubbed with dust from a sun-dried granary or with wheat flour, and bathed in lukewarm water. The new-born child is often placed on a winnowing fan, and sometimes upon a bed of rice. This is afterwards given to the midwife. The

[1] *Punjab Notes and Queries*, III. 188.

mother receives a warm bath. Sometimes the mother is not bathed until the sixth day, although she may receive a partial bath at a previous time set by the Brahman. The cord and placenta are buried in the house near the door to prevent their coming into the possession of an animal, or of an evil spirit, or of a magician; and over this spot in the house a fire is kept burning for six or more days. Some hide the cord in the house. The falling of the scab of the cord is watched with great care, and the particle is disposed of cautiously; most likely it is buried inside the house, lest it come into the possession of a *bhūt*, of a woman, or of a wizard. If a woman eat it, the child will die, but she will obtain children. If a wizard, or a witch, get possession of it, the child is sure to be ruled by their spells. If an evil spirit get it, the child will be possessed.

The announcement of the birth of a child is made by the midwife, or by a barber woman, or by a female relative, who does so by going to the house of the headman of the village and to the relatives of the family, and making a mark (*swastika*) on their doors with cow-dung. She receives a small fee for this. She also makes a mark on a shrine to Sītalā. If the child is born on an unauspicious day, a Brahman is called to perform a fire-sacrifice. Wood from thirty-six different trees is brought in for the purpose. The father sits in front of the fire during the ceremony. A cup of *sarṣoṁ* (mustard) oil is placed in front of the father, and the child is placed on his shoulder, so that the father may see his face reflected in the oil. After this service the father may look into his child's face. If no unfavourable conditions appear at birth, he may look at the child at once.

On the first day after a birth, a Brahman is consulted. He inquires in what direction the mother lay; how many women were present; and asks other similar questions, concerning the birth of the child. He then casts a horoscope, gives the name, and fixes a day for the purifying bath.

Strict seclusion is practised for an indefinite period, during which no one but the midwife and an old woman of the family are allowed in the lying-in room. During the six to fourteen days of her impurity, the mother is attended by these women only.

The midwife receives a wage of four pice when she cuts the cord, and four pice and some grain, usually barley, when she washes the baby. Besides this, she expects such wages and presents as the father may choose to give. Nowadays the fees are being increased, and in the cities the services of the midwife are fairly expensive.

After delivery, before the mother is given anything to eat, a quantity of gur is offered to the sainted dead. In the Central Provinces[1] the mother receives food neither on the day of delivery nor on the next two succeeding days; but, usually, after the mother and child have been bathed, the mother receives a special kind of food. This is a gruel made of a mixture of spices, gur, and oil. The food given to the mother, in the Central Provinces,[2] consists of a concoction of ginger, roots of *orai* or *khaskhas* grass, areca nut, coriander, turmeric, and other hot substances, and sometimes a cake of linseed or sesamum.

If her family is well-to-do, the mother will receive a helping of this gruel several times a day for twelve days. Besides this, she may receive milk two or three times a day. Food consisting of turmeric and ginger cooked in oil is served, usually from the sixth day. She receives ordinary food on the second day, or after six days, or after twelve days, according to the financial circumstances of the family.

The child is put to the breast on the third day, unless a Brahman[3] orders that this be done sooner. In some places before being put to the breast, the child is given a decoction made by boiling some roots in calf's urine.[4]

The child is not clothed for four or five days, and then the swaddling-clothes used should be borrowed from another person's house, or brought by relatives.

During the first six days the mother and child are never left alone, and someone is on guard every night lest some evil spirit obtain an opportunity to do harm. During this whole time the mother wears an iron ring, or an iron

[1] Russell, *Tribes and Castes of the Central Provinces of India*, Vol. II. p. 413. [2] *Ibid.*, Vol. II. p. 413. [3] *Ibid.*, Vol. II. p. 413. [4] *Ibid.*, Vol. II. p. 413.

instrument of some kind is kept under her pillow. The mother and child rest in the bed during the whole period of impurity, and the iron instrument with which the cord was cut is kept near the mother. The midwife, the mother, and the babe are considered unclean or tabu, and are not allowed to touch the food of the others.

On the night before the sixth day, the whole household sits up and watches over the child; for, on that day, his destiny is determined, especially as to immunity from small-pox and other dangerous diseases. He is carefully fed; for, if he go hungry then, he will be stingy all his life. This day of purification is called the *chhaṭṭhī*, and the ceremonies should be performed on the sixth day after delivery. Chhátthi or Shasti is the guardian goddess of children, who protects them from infantile diseases. Until they attain to maturity, children are supposed to be under her special care. She is regularly worshipped by women; and, when children are ill, her aid is invoked to effect the recovery. At this time her worship is especially efficacious in preventing lockjaw, a disease which not infrequently attacks infants about this time.

During the sixth night, the women do not sleep, but keep up singing and music, the beating of drums, and noise generally. They take special pains to protect the lamp which is burning, lest a peculiar insect *(janua)* put it out. If this should happen the child would die. Shasti is worshipped in the following way. On the wall, on both sides of the door, a square of cow-dung is made, in which one or seven broom-splints are fixed. This figure is called Shasti, and to it the women offer cakes of barley-flour and rice boiled in sugar. The child is now anointed with oil and lamp-black is put around its eyes. It is clothed and placed before the image; or the woman worships the image. The cakes are then presented on leaf platters and eaten by the menials of the family. *Halwā* is offered and sent to relatives. On this day the legs of the bed are worshipped.

Other precautions are taken against disease. The child is sometimes branded[1] on the stomach on the sixth

[1] Russell, *Tribes and Castes of the Central Provinces of India*, Vol. II. p. 413. The practice is common in other places.

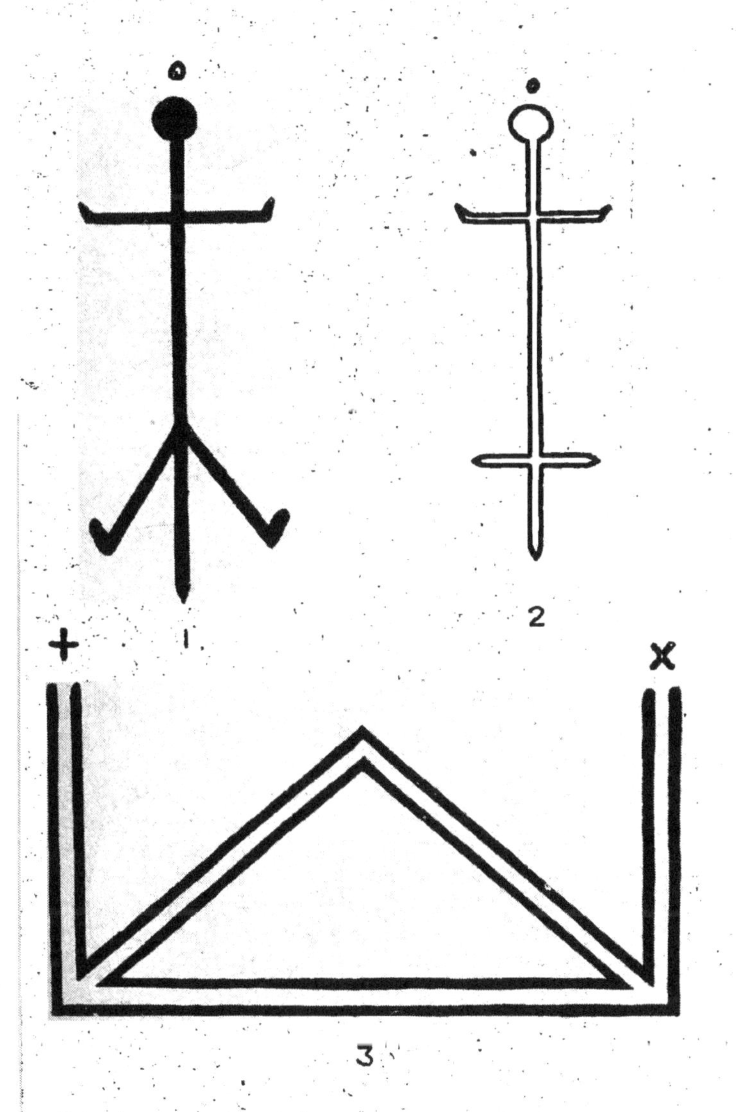

PENCIL DRAWINGS OF SHASTI (1 AND 2) AND OF SALONA (3)

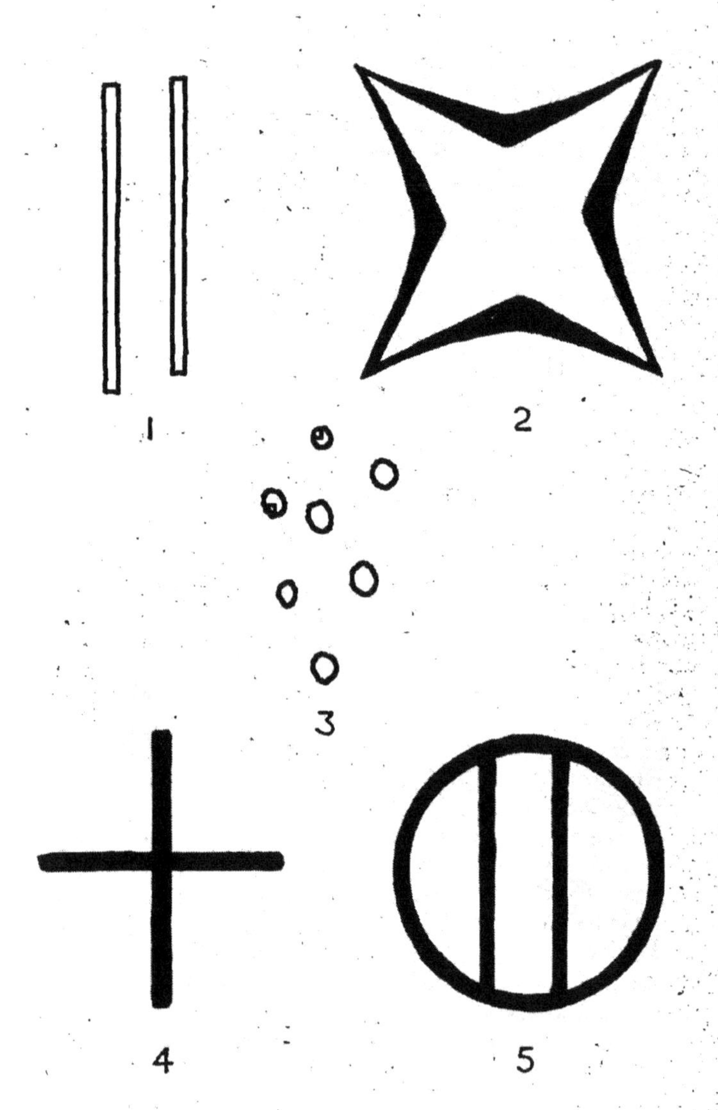

PENCIL DRAWINGS OF ABDOMINAL BRAND MARKS

(A little under actual size.)

day, or on the day when it is named; occasionally twenty or more burns are made on the abdomen with the point of a sickle to prevent the child's catching cold; castor oil is rubbed on him to prevent convulsions and lung trouble; and sometimes he is held in the smoke of the fire.

If at any time there be suspicion of the influence of the evil eye, a wave ceremony is performed. Mustard and dil seeds, or bran and salt, are waved around the mother's head and then thrown into a vessel containing fire. When all is consumed the vessel is upset, and the mother breaks it with her foot. She then sits with grain in her hand, while the household brass-tray is beaten, and the midwife throws the child into the air. Sometimes the baby is weighed in grain, which is then given to the priest or to the midwife. If they feel that the trouble is due to the influence of *Jakhiya*, the ear of a pig is cut and the blood is put on the forehead of mother and child.

After the worship of Shasti, the mother and child are bathed. When the water for this purpose is heated, ajwain and nim leaves are thrown into it. The mother squats on a plank during the bath. Under the plank a pestle, or a plough-beam is placed; or, if neither is at hand, a nail is driven into the ground under the plank. A cleansing draught, consisting of Ganges water and calf's urine, is then given to the mother. Sometimes the cleansing draught is composed of the five products of the cow[1] together with Ganges water. First a little of this mixture is sprinkled about, and then the remainder is administered. Afterwards, when the Brahman directs, the Shasti marks are removed from the walls of the house, taken to a well, sprinkled with water and left.

Besides the rites performed on the sixth day, similar ceremonies are carried out on the tenth, eleventh, or fourteenth day after birth: but more often on the twelfth. These are the final purificatory rites, after which the mother and child are considered clean. The house is then thoroughly swept and cleansed, and the room is sometimes

[1] *Pañchgavya.*

līped. The fire is removed from the lying-in-room, but afterwards is lighted again for three or five days. After the room has been līped, incense consisting of onions, garlic, red pepper and bran is burned, or an old shoe may be cast into the fire. The pice which were offered to the godlings, when the birth-pains began, are now spent for *batāśās*,[1] which are distributed in the name of the child; or gur may be given to the women who have helped. The husband's younger brother, or a sister, or the midwife, receives a gift of gur, and then takes the mother out of doors. Afterwards the mother takes grain on a brass platter to offer to the well. What is left of the grain is then given to the sweeper. She now bakes five loaves of bread and prepares a gruel, such as she received during the first days after her child was born, puts them in five places in the house, sprinkles water over them with her hand, and distributes them. After this she resumes her usual avocations.

Frequently, on the twelfth day, a black goat is offered to *Kālī Devī*, and a fire is lighted in her name. A feast (the *Dasatan*) is held. The father entertains his friends (the biradari), and the parents or the brothers of the mother send a coat and a yellow loin-cloth for her, and a red-cap for the baby. Sometimes they send sweets, *sataura*,[2] or *achhwānī*.[3] Feasts are often held on the seventh, tenth, and fourteenth days after the birth of a child.

When the baby is six months old it is fed with grain (in the form of *khīr*, rice cooked in milk) for the first time, and a feast is given. It is a day of rejoicing. Sometimes a Brahman is called in on this day to announce a name for the child, although the name is often given on the day of birth, and sometimes on the day of the chhatthi ceremony. The name given is often that of the day of the week on which the child was born; and, if he was born at the time of some religious festival, he may receive a name referring to that. The name usually given is one pleasing to the parents. Many give the child two names. The one

[1] A kind of sweetmeat.
[2] Somf made from ginger, spices and gur.
[3] Ginger, *ajwain*, etc.

obtained from the Brahman is kept secret for two reasons: first, because the child is thereby preserved from the magician's art and from evil influences generally; and, second, because he is more likely to be passed over by the angel of death. The second name is given by the parents, or by some old person of the family, and is the one commonly used. A feast is held and offerings are made to the sainted dead.

The sickness or death of either the mother or the child s attributed to the influence of evil spirits or of the evil eye. In case of illness, a wizard is called in to identify the evil spirit and to give directions as to what should be done to appease the demon. In former times, children dying at birth or in infancy were buried near the door, either in the floor or in the wall, so that the spirit might re-enter the mother's womb. In some places in the Central Provinces a stillborn child, or one dying before the sixth day, is placed n an earthen pot and buried in the court-yard, or under the doorway, and no funeral feast is held. Two ends are secured: witchcraft is forestalled and (they believe) another child will be born in the home. Occasionally, when the children of the family die one by one, a dying child is buried while still alive, so that the demon that besets the family may be buried with it. Usually, stillborn children are buried or cast into a river. The bodies of children over five years of age are cremated, except that the body of an unmarried child is not burned. A mother dying in child-birth becomes a *Churel*. Nails are driven into her finger nails and toenails, and powdered chillies are put into her eyes. Sometimes, when death occurs within ten days of delivery, a nail is driven into the door-post immediately after the corpse is taken out of the house. These are devices to prevent the return of the ghost to her former home.

Some peculiar superstitions prevail about certain irregularities at birth, and later. When a breech case occurs, it is believed that one parent will die soon, or that the child is likely to be killed by lightning; but, on the other hand, a person who suffers from backache may be cured if his back be touched with the feet of a child born thus. If the baby is born with teeth, it is believed that some crime will overtake the family, or that someone will die. To avert calamity word is sent to the maternal grandparents to send silver teeth. When the maternal uncle brings the teeth, he goes to the back of the house and throws them over the building so that they will fall at the door. In the olden time steps were taken to destroy such a child, for it was said that a cannibal (*rākshas*) had been born. Sometimes, when a child is born with protruding teeth, these are broken. If the upper teeth come through first, it is believed that some near relative on the mother's side will die in short time. Making a baby sleep towards the foot of a charpai tends to make the upper teeth appear first. Up to the sixth month no child should be lifted above one's head, lest calamity ensue; for an evil spirit may secure an opportunity to do harm. When a child cries a good deal they believe that it is likely to die soon, and, as a preventive, they pierce the nose. There are a variety of opinions as to whether twins are auspicious or not. If they be of opposite sex, the general feeling is that they are unlucky, but if both be of the same sex, their birth is fortunate.

There is no special ceremony at the time of puberty, and, therefore, no proper initiatory rite. Some say that after the purificatory ceremonies have been performed and his hair has been cut, a boy may be considered a member of the caste. Others maintain that from the time that the milk-teeth fall out, or from about the eighth year, he may be considered a Chamar. Others say that the marriage is the initiatory rite. Still others say that until a boy's ears are bored he may not join in such social festivities as smoking the huqqa. Usually, when a child is from five to seven years of age, his ears are pierced. Sometimes this is done at birth, or soon after. If he grow up with his ears unbored, he usually pierces them himself. A boy should not marry before his ears have been pierced. When a boy obtains recognition as a member of the biradari, he must conform to the social usages of his caste.

The case of girls is considered much more carefully. The first signs of puberty (*siyānapan*) are watched for most seriously, especially by the mother; and when these appear the girl is kept in a dark corner of the house. She will try to hide herself and to keep away from her friends and neighbours. She leaves her hair unkempt. This is a regular custom. At the first appearing of the menses, she must keep out of the sight of men; and she is secluded for four days, during which time no one touches her, not even her sisters, and she must not touch the food nor the cooking-vessels. Some say that she must not touch the thatch, nor trees, nor plants; that she must avoid the shadows of other persons; that she should carry a knife; and that she must not look upon the sun, a cat, or a crow, nor into the sky. Her food should consist of things prepared with sugar, curds, and tamarinds. She must not touch salt. On the fourth or the fifth day all of her clothes, and such clothing as she has touched, are washed. Then, accompanied by women, she goes to the village tank to bathe. On the way back she steps over a pestle. The seclusion enforced all this time is due to the superstitious fear of menstrual blood; the girl is tabu.

Adoption is effected in the following manner: After the panchayet has agreed to the proposal, the parents give the boy to those wishing to adopt him, with words about as follows: "You are my son by a deed of evil (*pāp*); now you are the son of so-and-so by a virtuous act (*dharm*)."[1] As the boy is accepted, members of the caste sprinkle rice over him; and then his foster-parents give a feast. All rights are made over to the new guardians, who are nearly always relatives of the boy. A gift, sometimes amounting to ten rupees, is made to pay the expenses of a feast for the biradari and for liquor.

[1] Crooke, *Tribes and Castes of the North-Western Provinces and Oudh*, Vol. II. p. 179.

CHAPTER IV

DOMESTIC CUSTOMS: MARRIAGE

IN making marriage arrangements Chamars, with some exceptions, do not employ barbers unless they be barbers of their own caste. When the parents decide, after the men of the family have talked it over, that a marriage should be arranged for a son, or for a daughter, they look about for a suitable mate for the child. When a desirable companion has been discovered, a go-between (*aguā, bichwānī, bichauniyā bichwāī*) is appointed to carry on negotiations between the two families, and to make preliminary examinations as to the physical fitness of the two children. He makes a report to each family concerned. This agent may be a relative or a friend of either party.

After the preliminary inquiries (*bāt chīt*) by the go-between, or match-maker, have been reported, the fathers of the two parties make similar inquiries themselves. At this time the family pedigrees and the gots are gone into. If an agreement is reached, a Brahman is consulted to ascertain whether the arrangements made are auspicious, and to fix the time for the betrothal proper. The girl's father then gives a rupee, or some such amouut, as earnest-money to the father of the boy. This amount is sometimes placed in the boy's hand. Refreshments consisting of crude country liquor are then served; in some cases this is paid for by the boy's representatives. Sometimes sugar is distributed, and the party is entertained.

The betrothal (*sagāī, mamgnī, barekhī, barachha*) follows. The girl's father, with male relatives and friends, goes to the boy's home to make arrangements. He then gives a rupee, and makes a mark (tika) on the boy's forehead with rice and curds, or turmeric, saying, "I have given you my daughter." This rupee is the sign (*niśānī*) that the engagement has been made. [In some places four pice of the old coinage (*kachchā paisā*) are given and in others a rupee and a loin-cloth.][1] [Some give two and a-half yards of cloth and a sum of from five to twenty-five rupees. After the placing of the tika upon the boy's forehead, the bichauniya says, "This union which the elders have made may Parmeshwar cause to turn out well." The boy then stands up with the cloth and the rupees in his hands and salutes his elders, and they in turn say, "May you live long." The boy then takes the cloth and rupees to his mother, who is in the house behind with other women who are singing and beating drums. The boy's father gives the company sweets and liquor. The women of the clan join in music and singing. The boy is then seated in the east, or west, as the Brahman may direct, and receives the niśani from the hands of the priest (Brahman), or from some relative of the girl. The men repair to the village liquor-shop, or liquor is brought. Sometimes the boy's father distributes sugar to the head-men, and to the Chamars of the village, as proof of the maṁgni; and occasionally he feeds the clan. That night the boy's father gives a dinner (*dāwat*) to the girl's father and friends; but immediately after the feast each of those who have come from the girl's house gives a rupee or two to the boy in payment for the food. In the morning they return home, giving the boy another rupee before they start.]

[In some parts of the United Provinces the regular betrothal takes place at the village liquor-shop. On a day agreed upon by the parties concerned, the fathers of the young pair meet and exchange cups of liquor five or seven times. At the last round the father of the prospective bride puts into the cup from which the other drinks, a rupee. This ceremony, called *piyālā* (cup), is the binding element in the betrothal. The persons present now

[1] This account of the marriage customs is based primarily upon reports from Jaiswars. The more detailed alternate practices of other sub-castes are placed in square brackets.

proceed to pass liquor around freely. Money is collected on the spot by those belonging to the girl's party, and this is supplemented by donations from the others.]

At the time of the betrothal the girl may be but six months old, or even younger. In earlier times tentative arrangements were sometimes made even before the birth of a child.

The mamgni is all but irrevocable. Either party, however, may break an engagement, with the consent of the council (panchayat), by paying twenty-two rupees, or some such sum, part of which goes to the chaudhari, and the rest to the other party to the mamgni. The causes for which a betrothal may be broken are strife between the families concerned, or the discovery of an incurable disease or of an infirmity in either the boy or the girl.

Some time after the mamgni has been performed, usually when the milk-teeth fall out, or the girl is about eight years of age, her parents send a letter fixing the date for the marriage. This is followed by gifts of nine yards of cloth and two and a-half or five or ten seers of grain, two betel nuts, some grains of rice dyed yellow, five pieces of turmeric and a sheet of paper with the order of ceremonies written upon it. When these things are brought the boy's father calls his relatives and the chief men of the biradari together and announces that the *lagan* has come. The groom is then called and the paper together with a rupee are placed in his hands, gur is distributed, and a feast is given to the friends. The boy's father, with male relatives [in some places a relative and the go-between], then goes to the prospective bride's home bearing gifts in shallow baskets. These presents consist of five seers of rice, five seers of sugar, mango and *tilli* wood for the fire-sacrifice, and sacred grass, betel nut, haldi, *sindūr*, coarse white thread and two loin-cloths each six or seven yards in length. When these things have been received the chaudhari is called in to open the baskets and show the presents. When evening comes on, the girl's father gives a rupee to the boy's father in order that the latter may be willing to share in the feast which is being prepared.

For this dinner rice, *dāl*, meat, and *baṛā phulauṛī* and special oil-cakes (*dusṭī*) are served. The cakes are baked double. There is some drinking. The well-to-do have dancing exhibitions (*nāch*) also. The women and girls join in singing special obscene songs, in which the abuse is directed towards the visiting party. The next morning a Brahman is called in to fix the time for the marriage ceremonies. He draws his figures on the ground, using wheat or barley flour, and proceeds to announce the times for the various parts of the wedding ceremony, such as the cutting of the wood, the cleaning and grinding of the grains for the feasts and ceremonies, the invitations (*neotā*), the *maṭ maṁgrā* ceremony, the erection of the marriage pavilion, the anointing of the bride and groom, the pre-wedding feast (*bhaktawān*), and the *bhāṁwar*, or phera. He then prepares a fire-sacrifice, using the wood brought by the boy's relatives. Into the sacred fire ghi, sandal wood and incense are offered by the girl, as the priest directs, and she calls upon the Ganges and other gods as he names them to her. Then the Brahman ties a *kaṁgnā*, which he has made on the spot, around the right wrist of the girl. The kaṁgna is made of chaff and rye and anise-seed (*sauṁf*) bound in cloth with yellow thread, *e.g.*, thread coloured with haldi. Into this kaṁgna an iron ring is tied. The Brahman then puts sacred grass (*kuś*) into the hands of the girl, and the boy's father places a gift in her hands, which she in turn puts into the square which was drawn for the sacrifice, and sprinkles water over it. The priest takes the gift and then leaves the house. After another meal the visitors depart.

On this day announcements of the dates for the wedding festivities are made. Two lumps of coarse sugar are sent to each family which is to be invited to the wedding. The bearer of the gur announces the dates. The woman who bears the invitation sits outside the mahalla and sends word that she has come to announce the wedding. The women come out, singing, to receive her and bring her in. She then gives the invitation and distributes the sugar.

[An alternative practice is as follows: A Brahman is summoned to the door of the girl's house. Early that morning the father and the girl have both bathed. A place has been lîped in front of the house and a chauk traced on it with flour. The *paṭā* on which the girl sits is placed over this. The Brahman sits at the right-hand and the father at the left-hand corner. And a new pot (*korā*) of water is placed outside the square in front of the girl. The people of the village, who have been summoned by the barber, are present. The Brahman recites some *mantras* and then proceeds to write the lagan, which is a document fixing an auspicious day for the *sādī*. He then puts five betel nuts, a handful of rice, a piece of turmeric and fifteen rupees on to the lagan, and rolls it up and ties it with string. Five suits of clothes are also given, and the girl's paternal uncle (*chachā*) adds from two to five rupees, and her father's sister's husband (*phūphā*) the same sum. The Brahman's fee is one rupee and half a seer of *chana* from the father of the girl, and one rupee sent to him by the father of the boy, when the lagan has been delivered. The lagan is sent to the boy's house by the girl's father, by the hand of men whom he, in order that they might get their clothes washed, had warned eight or ten days before that this service would be expected of them. The girl's father also goes himself and her phupha and the bichauniya and three or four others. With the lagan are sent as many rupees as were given at the mamgni and a suit of clothes for the mother of the boy. When the lagan arrives at the boy's village, his father sends around a bhamgi (sweeper) with a drum (*dhol*) to announce the arrival and summon the villagers. As soon as they assemble an announcement is made of the value of the gifts sent, and the boy receives them with much the same ceremony as in the case of the mamgni and takes them to his mother. As soon as this is over, the father of the boy produces five *behlis* of gur and one to three rupees' worth of batasas. Half a behli is given to those who have come from the girl's house; they take it to the *chaupāl* (a public hall), where they eat it and smoke. The people belonging to the village eat the rest of the sweets at the boy's house. That night those who have come from the girl's house are entertained, but give two rupees in payment for their food, and place two rupees more in the boy's hands when they depart next day. All night long the women sing, while two or three of them in turn sit at one side and cook *pūā* (*āṭā*, flour, cooked in oil) in a large iron vessel (*kaṛhāī*). In the morning they go into the village and distribute it from house to house.]

On the days appointed, the following marriage preliminaries are carried out in both homes.

On the day set for the *maṭ maṁgrā*, or *maṭ koṛ*, or magic earth ceremony, relatives and friends come with gifts of grains, wood, clothing, oil and sweets. They come singing and beating drums. Then the women, including the mother of the bride, or groom, take a brass tray, or a basket, with sugar, pulse or gram (chana) and a one-wicked lamp (*chirāg*) and go in procession to the village clay-pit. They are preceded by a Chamar beating a drum. The women sing as they go. Then they worship the drum, marking it with red-lead (tika). They mark seven, or five, places about the pit with mustard-oil and red-lead (sindur). Seven, or five, women are then chosen, each of whom takes a clod of earth from one of the places so marked, and puts it into a basket. They then distribute the sugar amongst themselves, after which the mother carries home the seven clods of clay. From this earth is made the fireplace for the cooking of the marriage feast; and in some places the family grindstone is repaired from some of the same clay. In some places the earth is brought without any ceremony. On this day the women go to the potter's house, with presents of grain, worship his wheel, and get the earthen pots used for furnishing the marriage pavilion and for use in the house. In this connection *Būṛhā Bābā* is worshipped. In some places a special pot (*kalsā*) is ornamented and set in the thatch.

The *māṁḍhā*, *māṁṛo*, or marriage pavilion, is erected on the day that the magic earth is brought home. Sometimes the maṁdha is set up on the day when the barat comes. A grass rope is made by a maternal uncle and hung over the doorway of the house, and sometimes a winnowing fan is hung against a doorpost. In the courtyard, in front of the house, four (in the hill country some use nine poles of the *siddh* tree) bamboo posts are set up and a thatch is built over them. This pavilion is large enough to seat from twenty-five to thirty persons. In some places, two green bamboos are set up to support an awning of thatch which is attached to the house above the door, and occasionally but one post is used. Sometimes five plows are planted to form the shed. On each side of the door earthen vessels of water are set. Into one rice, and into the other pulse, is thrown. Mango leaves are also used. Earthen lids are put upon both vessels. The necks of the jars are bound with yellow and red threads, and each is tied to a bamboo post with a rope of grass into which mango leaves are bound. In the centre of the pavilion many things are set up, but local custom determines which of these articles shall be used. A green bamboo and a plow-beam are set up by five men. Under the bamboo two pice, two pieces of turmeric, two betel nuts and rice are buried. The plow-beam is worshipped as it is set up, and the maternal aunt places her hand-impression upon the beam five times in a paste of ground haldi and rice. She also puts her hand-print upon the backs of the five men who set up the pavilion. Mango leaves and a kamgna are bound upon the plow-beam. In some places a small earthen pot, bound with grass, is attached to the beam. This pot is ornamented with crossed lines made with rice flour and turmeric. Five marks are made upon the beam with red-lead, and a brass pot, or an earthen one, is placed beside the beam. The log used to break the clods in the plowed field is often set up also. A lamp is bound to this log. In many places a branch of the *dḥāk* tree is erected. Against the supports of the maṁdha, a *sil* and *baṭṭā*, the stones used for grinding spices, are placed. Sometimes a rolling-pin (*belan*) is used. Along with the plow-beam a marriage-ceremony pole (*sūgā*) is often set up. This is made of mango wood. To it are attached branches having rude wooden figures of parrots perched upon them. "After the wedding

MARRIAGE POLE (SUGĀ)

KOHBAR

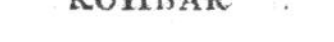

POTS SET IN THE ROOF AT TIME OF MARRIAGE

BLACKENED POT WITH WHITE SPOTS UPON IT SET UP IN A FIELD

POTS AROUND A BAMBOO ON A MUD PLATFORM NEAR A VILLAGE

CHAMUNDA'S PLATFORM

there is a general scramble for the wooden parrots, but the pole is carefully kept for a year." In the pavilion an earthen vessel (kalsa) is placed. This is decorated with lines of cow-dung horizontally and vertically drawn. In these lines grains of barley are stuck. Some place a pot near the plow-beam, partly fill it with water, and then put in oil and a wick and light it. Or a cover is placed over the vessel and a wick lighted in this. Green mango-leaves are inserted between the vessel and the cover. The poles of the pavilion are hung with coarse white threads in which mango-leaves are bound. The gobar remaining from plastering the floor of the maṁdha, or a piece of cow-dung, is left in the shed. The pavilion at the bride's house only is thatched. The roof is made of *sarpat*, a coarse grass. [Some Jatiyas set up a plow-beam in the courtyard at the bride's house, put a pot, marked with white lines, on the top of the beam, and stretch a *śāmiyāna*, or canvas, on this. They do not set up a bamboo with the plow-beam, nor is the beam worshipped or marked with red-lead. The pot on the plow-beam represents the head of the Babrabān, a ṛishī whom Kṛishṇa feared, and consequently slew. The ṛishī said, "I have not seen the fight"; so Kṛishṇa said, "You shall be at all weddings, and see them at the bride's house." In the groom's house seven earthen platters with puris sandwiched in, and a hole bored through them, are hung up in the pavilion. A rope is put through the hole and the platters are tied to a bamboo, which is set up instead of the plow-beam. Two pice are buried beneath the bamboo. No canvas is spread.]

On the evening of the day when the magic earth is brought home, and the pavilion is set up, the Brahman is called to the girl's house. He prepares a fire-sacrifice, this time offering in the fire the things which had been brought from the home of the groom for the anointing (*ubṭan*) of the girl for the wedding. He mixes mustard oil with turmeric and barley flour (one half of the flour is parched and the other half is not) for the anointing, and touches her forehead and shoulders with the mixture, using sacred grass (*dūb*) for the purpose. This fire-sacrifice takes place in the maṁdha.

In the courtyard a square is covered with cow-dung and marked out with crossed and recrossed lines of slaked lime. Upon this a low four-legged stool (*pīrḥā*) or a plank (*paṭrā*) is placed. The legs of the stool are bound with coarse coloured thread. An earthen vessel of water is brought. This also is bound with coloured threads. Then a small vessel for pouring water, and then the vessel containing the anointing mixture are brought. These also are decorated with threads. The bride is now led out and seated upon the stool. In some places she worships the goddess in the courtyard, before taking her place here. In other places, seven women, seven times each, sprinkle her head with the mixture with dub grass. Often the bride's feet are washed, ceremonially, with water, and water is poured around her on the ground. Then water is poured over her and she is rubbed again. (This may be done only on the day of wedding, at the time of the last anointing.) Then the women-folk rub the girl thoroughly from head to foot with the mixture of oil and turmeric. The Brahman then puts rice and gur in the bride's hands. Five unmarried girls sit down around the bride, and each of them in turn touches her toes, knees, and forehead, and then they kiss their hands. The mother then, taking a *chādar*, covers herself and her daughter with it, and leads the girl to a specially prepared space in the house; or she is carried into the house. The gur and rice are then put down beneath a figure drawn on the wall, and the five girls come and distribute the gur amongst themselves. The rice that is deposited here is given to the maternal uncle after the wedding. This special place is prepared by a sister, or a maternal aunt of the girl. On the wall above the place, a rectangle known as the *kohbar* is drawn in red and white, with circles at the four corners. Within this figure, pictures of horses, elephants, birds or other objects are drawn in colours (red, yellow, green, white).

A similar anointing ceremony is performed for the boy. In the ceremony unmarried girls assist just as for the bride.

Six more times before the wedding each is anointed and led to the appointed place in the house, with rice and gur in their hands. The last anointing is usually performed on the day of wedding, but in some places these preliminary preparations are completed a number of days beforehand. In this case the anointing with oil takes place each morning. The women sing each night beginning on the day of the lagan. In other places the anointing begins, for the boy five days and for the girl three days before the wedding. The so-called "oil" for the anointing is composed of wheat, or barley flour, haldi, and water. With the present that is sent to the bride for the ubtan the balls formed from the "oil" as it was rubbed over the boy are included. Each day, after the anointing (for both boy and girl) a brass platter is brought and on it sugar, rice, haldi and a lamp containing ghi are placed. The platter is then waved before the candidate. Then five sweet cakes are placed in the child's lap and he throws them over his right shoulder. The cakes are caught by the sisters of the child, or by his father's sisters, in their clothes. Then the child is carried into the house. At this time the girl's hair is carefully unbraided by young girls.

Some time during the day when the magic earth ceremony is performed, seven women, who belong to the family, each take a pestle (*mūsal*) upon which coarse, unfinished red and yellow threads are wound, and with them they husk the rice used in the evening ceremonies. Sometimes but one pestle in used.

On the day that the sacred earth is brought, a fireplace (*chūlhā*), open on four sides, is made. On the day before the wedding, this fireplace is set up in the pavilion. When the chulha for the wedding-feast is ready to light, four women lead the bride's (or groom's) sister's husband to the fireplace, where he offers ghi and sugar and then lights it. The women sing while this act is being performed and afterwards give the man four pice. Upon this fireplace a feast is cooked consisting of rice, pulse (dal), fritters (puri) made of wheat flour cooked in oil or ghi, a preparation of curds and ground gram (*kaṛhī phulaurī*) and cakes of gram flour prepared in oil. This food is then served on five plates made of leaves. A pot, or a lamp, is now offered in the pavilion. This is the sacrifice of bhaktāwan. Then one of the plates of food is given to the girl (or boy) and the other four are given to the parents and other near relatives, the rest of the food being served to the remainder of the company. [In other places the ceremony is as follows: Upon the new chulha, fritters (puri) are made. The first ten fritters, together with rice, sugar and a lamp with four wicks, are placed upon a brass tray and waved in a circle before the face of the boy (or girl). The tray is then put into the hands of the boy (or girl). The relatives who participate in this ceremony take the rice and sugar from the tray and throw it to the right and left. The boy (or girl) is then taken into the house, where some of the puris are given to him. At the same time a portion of the ten cakes is distributed among those present. A feast follows. While the dinner is in progress a quantity of liquor is put into a small earthen cup (*kulhiyā*) which is set in a hole in the floor of the pavilion, and the father then removes the cup with his teeth and drinks the liquor. There is some drinking during the meal. In the evening there is more feasting, and during the night a good deal of drinking is indulged in. As soon as the girl has finished her dinner, her mother places her hands over her eyes, and leads her to the village dung-hill, where the girl buries her plate.]

[In some places, on the night before the groom's procession leaves for the wedding, and just before dinner, the women take the groom into the house, call a potter, light a lamp filled with ghi, put fire before the lamp and then empty the ghi out of the lamp into the fire. The fire blazes up suddenly, and they say that Burha Baba has come and that he is well pleased. Gifts are then made to the potter, and he binds a kaṁgna on the wrist of the groom (or bride). The wedding-feast follows. On this day, up to the time of the dinner, the parents do not drink water, and they fast[1] until the dinner is served on the wedding-day.]

[1] Those who fast are the maternal uncle, the father, the mother the brother, the paternal uncle, and the mother's sister's husband.

On the morning of the wedding-day the girl (or boy) may receive the last anointing. The father's sister's husband, called negi, hollows out a place in the court-yard of the house. Here the boy is made to stand while he is bathed. The father makes a small present to the negi. Beside the boy is set a winnowing-fan (*sūp*), in which his relatives and friends place their offerings, sugar, rice and money. This is all taken by the negi. The boy is then bathed. The first water that is poured over the boy's head is caught in an earthen vessel. This water is preserved to be taken to the bride's home for use in her preparatory ablutions. After the boy has dressed himself, he is led to the pavilion and seated in a square already prepared by his uncle's wife. This woman then puts lampblack (*kājal*) into his eyes, and marks his forehead and temples with a paste made of ground rice. Then the negi pretends to cut the boy's finger nails and toenails. While this is going on the relatives drop coins into a brass pot (*thālī*) of water which has been placed in front of the boy. A crown (*maur, tāj*) is placed on the groom's head by a male relative. This is worn during the succeeding ceremonies until the marriage is completed. The women sing during this and during most of the preliminary ceremonies. Puris are prepared in the boy's home. Seven women, each taking two puris, together with a piece of sugar, go seven times around the maṁdha, with the left side towards it, unwinding coarse white thread (*kukaṛī*) as they go. The boy and the mother sit in the maṁdha during this ceremony. After each round the boy takes a bite out of each puri. At the seventh round, his mother's brother puts bits of the puri and water from the pot in the maṁdha into the mother's mouth. She tries to gulp all this down. This is called *imlī ghoṁṭāī*. In some places the central fibre of a mango leaf is used in place of pieces of puri. Five girls take rice and sugar in their hands and touch the groom's feet, knees, forehead and temples, and then kiss these articles. The boy is led to the kohbar, under which he places the rice and sugar, and then the latter is distributed amongst the women. After the marriage the rice is given to the negi. The boy is then conducted to

6

the place where the men are to eat and drink before starting to the bride's house. After this meal an exhibition of dancing is given by Chamar men, specially called for this purpose. Some of the dancers are dressed as women. These exhibitions are not of an elevating nature. The performers accompany the groom's procession to the bride's house. The groom either walks in the procession, or is carried on someone's shoulder, or rides on a horse or in a *dolī*.

Before the groom's party leaves, the mother performs a wave ceremony. She first makes seven lamps (chirag) of flour, places them in a winnowing-fan, and waves it seven times about the boy's head. She then throws the lamps in seven directions. One of the dancers now seizes the fan and throws it over his head backwards. The fan is then taken into the house. She then waves a lota of water seven times about his head, pouring a little upon the ground each time. Likewise a pestle, a grinding-stone (batta), and his mother's chadar are waved about his head. Sometimes a wave ceremony is performed with a four-wicked lamp in a brass tray.

The mother then goes to the village well and sits down upon the curb, or even puts her feet over the edge of it. She does this with the pretence of destroying herself because her son will neglect and fail to support her after he is married. The boy then comes to the well, walks around it seven times, and marks it with his fingers with rice-flour and turmeric. He then takes his mother home, comforting her by saying that he will continue to take care of her, and that he will bring her a baṁdi laumdi to serve her and to wait upon her. [In some places the party goes to the well. There a puri is pierced with an iron rod. The boy looks into the well and then shakes the puri off into the water. He then returns from the well and, at home, takes his mother's breast.]

The groom now joins the procession. After they have paid their respects to the village godlings they start for the bride's home. He takes presents of clothes for the bride's male relatives (sometimes for the bride's sisters as well). The father is supposed to take a neck-ring (*haṁslī*) and wrist-, ankle-, and ear-rings of metal for the girl. Noisy music is a feature of the barat. With the marriage procession a special dance, sometimes obscene (*naṭwā nāch*), is performed by male members of the tribe, some of whom dress in women's clothes.

The wedding ceremonies are directed usually by some older relative, as the negi, but sometimes by a *mahant*, or by a Brahman. In some places the groom's father's sister's husband directs the ceremony.[1]

As the actual marriage always takes place at night, at an hour fixed by the Brahman, the barat is timed to reach the bride's village late in the day. The marriage procession stops a short distance from the village, and drums are beaten and horns are blown to announce its arrival. At this time, in some places, the girl's father, with some near relatives and friends, goes out to meet the groom, and the barat is led to a specially-prepared place called the *janwāṁs*. Then the groom is led to the maṁdha, or is carried there on someone's shoulders, that he may shake it. Or, when the groom arrives at the door of the bride's house, he is met by her mother, who performs a wave ceremony. She then places seven earthen saucers in her chadar and sits upon the ground. The groom is challenged to break all seven with a single kick, and is taunted as the eldest son of an old woman. He succeeds, however, in meeting the challenge. He is then returned to the place where the barat is waiting. A maternal uncle of the bride comes and washes the feet of five relatives of the groom. Or, this may be done before the groom's party is led to the resting-place. These men place their feet in a basin (thali) for this purpose. Then the negi brings gur and curds, with which he feeds five men, and after which he receives two *ānās* (*neg*). This man then brings and distributes cooked rice mixed with uncooked pulse (*uṛd kī dāl*). In the eating of the uncooked dal there is a symbolical test of strength. The basket in which this food was brought is now broken.

[1] Crooke, *Tribes and Castes of the North-Western Provinces and Oudh*, Vol. II, p. 181.

While the barat is waiting for the wedding ceremony, the female relatives of the bride sing obscene songs, in which the abuse is directed towards the groom, his relatives and friends. The women indulge in obscene and coarse language also; and cow-dung, mud and unclean things are thrown.

After the negi has carried his gifts to the bridal party, the girl receives her bridal bath. A place is hollowed out in the courtyard, just as was done for the boy. The negi brings the water that was preserved from the boy's bath, and his wife pours it over the bride's head. Then the women bathe the girl, and she is dressed in the wedding-clothes brought by the groom's party for the purpose. The mother then offers her breast. The maternal uncle's wife anoints the bride's eyes with lamp-black, and puts a ring (haṁsli) about her neck. The nail-paring and the imli ghoṁtai ceremonies are performed. The girl is then conducted to the marriage pavilion and seated in a specially-prepared place. The bride is now seen for the first time by the groom's party, but her face is veiled.

[In some parts of the country, a cock is brought and placed at the boy's feet. Sometimes its toes are cut off. Later it is offered to the sainted dead, and eaten.]

The actual wedding (*śādi, biyāh*) then takes place. The groom is brought by his paternal uncle into the pavilion and seated at the left-hand side of the bride. [Or they are seated facing each other, the bride's face being covered with a paper mask with seven broom-splints fixed in it over her forehead.] The negi now performs a fire-sacrifice, and then the bride's near relatives worship the feet of the pair. A new brass tray (thali), filled with water, is brought. Against the edge of this the boy's right large toe and the girl's left great toe are tied together. The parents of the bride now dip sacred grass (kuś) and the corners of their loin-cloths (dhoti) in the water, touch the great toes of the pair, and then their own foreheads, repeating the act seven times; and those who fast sip water from this foot-worship. Then presents, such as metal cooking-vessels, coins, and clothes, are given to the groom. This is the ceremony of giving away the bride, *kanyādān*,

Then the parents step aside, while the boy comes in front of the bride and marks her forehead seven times with sindur and places in her lap a small metal box containing the same kind of red powder. This marking is done with the thumb and little finger of the right hand, and the marks extend up into the parting of the hair. This is called *sindūrdān.* [Sometimes both the bride and groom have a red mark with a grain of rice in it imprinted upon their foreheads by their brothers-in-law. Then an unmarried girl is called. To her are given four anas, which she waves seven times over the bride's head.]

The binding part of the wedding ceremony (*bhāṁwar, pherā, biyāh, śādi*) follows. The preceding ceremonies are planned so that this takes place after midnight, even as late as four o'clock in the morning, at the hour announced as propitious by the Brahman.

Before the bride and groom were brought into the pavilion for the ceremonies just described, a square was drawn on the ground, in the marriage shed, with wheat and barley flour. In it diagonal and median lines were drawn. At the corners spoon-shaped decorations were made. In front of this chauk the bridal pair were seated. In this square a fire was lighted and in it offerings were made. During the ceremony which follows, the women sit in the pavilion and sing. The groom's maternal uncle also sits in the maṁdha and other relatives of the bride may sit in it. Others of the bridal party sit outside. Then the circumambulation (phera) is performed in the marriage pavilion as follows: First the fire is covered with an earthen saucer. Then the brother of the bride puts rice in seven places upon the sil, and some parched, unhulled rice in his loin-cloth, and takes his stand in the maṁdha. The bride draws her chadar over her face. The corner of her veil is tied to the groom's clothes and in the knot two copper coins (pice) are enclosed. The boy then leads in circling about the beam, or pole, so that his left hand is next to it, seven times. Or each leads three and a-half times about th[illegible]ole. Each time that the couple pass the bride's brot[illegible]e takes out of his loin-cloth, with a small round bas[illegible], a little of the parched rice, waves it over the heads of the couple once, and throws it on the ground (*dāl maunī*). At the same time, the groom throws away a pile of rice from the sil. Or, while going around the pole the seven broom-splints are removed, one at each round, from the bride's mask. Before going around the pole the pair exchange shoes. The couple now return to their seats, but exchange places. The groom's elder brother, or some other relative, now throws coarse silk and cotton threads, of red colour (*dhāg bhāṭ*), which are tied together, over the bride's head. Then rice and sugar are put into the hands of each of them, and five young girls touch the toes, knees, shoulders, and foreheads of groom and bride with sacred grass, and kiss it. Other things are sometimes placed in her lap, such as plantains, cocoanuts, mangos, a lamp, or a boy.

Immediately after the phera the couple are conducted by the women to the kohbar, where the bride's mother is sitting. Then they worship the threshold and eat together. The groom is stopped at the door by the bride's sister, who requests him to repeat a verse of something. This he refuses to do until he receives a present. Then he recites the verse, takes off his shoes, and enters the house. If he is silent, or too nervous to speak much, the bride's sister may, as a joke, steal his shoes while he is inside and hide them, in order to compel him to speak and say, "Where are my shoes?" Rice and sugar are placed before him, and then the bride's mother brings curds and sugar in a brass vessel, and the groom is required to partake of it, and is even bribed to do so. He takes a small portion, pretends to touch it to the bride's lips, and eats it. The knot is now untied, and the bride remains in the house, while the groom returns to the barat.

[In some parts of the country, the phera is performed a little differently. In the courtyard, in front of the marriage pavilion, a quadrangle about two feet square is marked out with barley flour. In each corner a bamboo peg is driven. Around this quadrangle, thread is wound, usually at the time of the phera. On one side of the chauk, but not in front of the pavilion, sits the boy's uncle on his mother's side (*māmū*), and on the opposite side of the quadrangle sits the girl's mamu. Their seats are short-legged stools (pirha, pata). Within the quadrangle a platform of magic earth, or of cow-dung, is made. Sometimes a plow-beam is set up in the square. A fire of dhak wood is lighted in an earthen vessel (kalsa), set in the middle of the enclosure and worshipped. Ghi is offered in the fire. A similar offering is made in the fire in the *deokurī*. The names of the couple and also of their fathers, grandfathers, and great-grandfathers on both sides are recited (*gotrā uchhānā*). About this quadrangle the couple walk. This is the binding part of the ceremony. The girl is blindfolded. A corner of her chadar is tied to the boy's clothes. Then they circle about the fire in the direction of the course of the sun, three and a-half times, the boy ahead, led by his father's brothers-in-law, and three and a-half times, the girl ahead, with her father's brothers-in-law leading. As they go, they wind unfinished thread about the pegs of the quadrangle. During this part of the ceremony the company throw rice upon the pair.]

[Then, in some places, a goat or a ram is sacrificed to *Parameshwarī Devī*. The flesh of the slaughtered animal is cooked for the marriage feast.]

After the phera follows the marriage feast. There is much drinking both at the place of the wedding and at the nearest liquor-shop, and much dancing and carousing continues until early morning.

An illustration of the coarse joking that takes place at this time is the following: The bride's mother dresses in men's clothes and, going to the groom's father, addresses him as "wife." The subject of the conversation is exceedingly vulgar and the result is a good deal of mirth.

The next day, or a little later, preparations for the departure of the bride and groom are begun. The couple, accompanied by the dancers of the groom's party, or the boy only, are taken by the bride's father to the village landlord. After the dancers have performed, the landlord makes a present (of from one to ten rupees in cash, or of cultivating rights) to the bride's father and then the party returns home.

The brass vessel used in the foot-worshipping ceremony of the day before is then placed in the maṁdha and the bride's relatives drop coins into it. These are collected by the bride's father and presented to the groom's father, in the brass tray. The bride is then prepared to go to the groom's home. And the bride's father says, "We have nothing else; we give you our daughter. May no harm come to her." Liquor and parched gram are passed, and afterwards a midday meal is served. Then the groom's father gives to the bride's father a gift for servants' expenses. The bride is then dressed in her wedding garments. The wife of the negi (her relative) ties the bride's veil to the clothes of the groom and then the bride's mother performs a wave ceremony about the heads of the pair. This is similar to that for the groom before he started for the wedding. During this wave ceremony the groom hangs on to the veil of his mother-in-law and does not release it until she makes him a present.

After the couple are seated in the conveyance, or are ready to start, the relatives of the groom return and shake the poles of the maṁdha, and pretend to untie the strings holding up the thatch. Then a little rice and sugar and two pice are put into the bride's lap. The couple then enter an ekka, or some other conveyance, a bahli, a doli or a *gāṛī*, and start for the husband's home. Or they go on foot.

[In other parts of the country the ceremony is different. In the early morning, before the barat starts back to the boy's home, a bed, given by the girl's father, is brought out into the courtyard. Upon this the bride and groom are seated, and then they are covered with a sheet (symbol of the consummation of marriage). The bride's father then gives the groom a rupee and other presents, such as vessels and clothes. Others also make presents at this time. Each one who makes a present puts a tika on the boy's forehead. The gifts are presented in the grain-sieve (sup), or in a basket, or on a tray, and are placed upon the bed. When the boy gets up, his brothers-in-law, or his father's brothers-in-law, take up the presents for him.

The groom then takes his place in the marriage pavilion, or in the janwaṁs with one or two relatives.

The bride's female relatives come from the house, and first wave pice in a circle before the groom, and then present them to him, salute him, and retire. Then the groom, after receiving a gift of about a rupee, unfastens one of the knots holding the roof of the maṁdha (*māṁḍhā-kulhāī*). The bride alone, or with the groom, then enters a conveyance of some sort. Thereupon her female relatives bring water from the house and wash her face. This is called *kuṁwārpan kā ūṭhānā*. She is no longer a child. Then they throw rice upon her, and again, as the procession starts, rice is thrown; and the boy's father throws money over the conveyance.] [After the barat has proceeded about a hundred yards, the two fathers embrace, having bared their breasts to do so, and the father of the bride gives seven rupees to the father of the groom and tells him that he will come after a few days to bring his daughter home.]

When the bridal-party reaches the groom's home, the bride worships the feet of her mother-in-law (*pāṁw pūjā*), and sometimes, his brothers and sisters worship her feet. The bride arrives with her face covered, and, as the women of the groom's house come to look at her, they make small offerings. She is then led to the kohbar, where she is seated and given a little food. This is the sign of admission into the clan. In some parts of the country, care is taken that the bride in no way touches the threshold as she enters. Often, in connection with the eating of her first meal in her husband's house, she has to step over a number of baskets.

A contest takes place at the kohbar. The necklace belonging to the bride is taken and thrown seven times at the square on the wall of the inner room and the couple struggle for it. In some places, a similar play is performed on the day following the arrival of the bride at her new home. A kaṁgna is bound on the right wrist of the boy, with seven knots. In the same manner another is bound on the girl's wrist. Then, with their left hands, they untie each other's kaṁgna and the most expeditious one is declared the winner. Then both kaṁgnas, together with a pice, are dropped into a vessel of water (or of whey, or of diluted milk) by the boy's sister's husband, or the father's sister's husband, three or seven times, and the two grab for the coin each time. The one who secures the coin a majority of the times is declared the winner. These tests are supposed to give some indication of who shall be the ruler of the house.

The next morning some food is prepared under the maṁdha. This is placed on five plates, made of leaves. Then the maṁdha is pulled down and taken in procession to the village tank, the bride and groom following, where water is sprinkled over it and it is buried. Then those who are present drink or wash their hands and faces in water taken in the earthen vessel which was placed in the maṁdha on the day when it was built. When they have returned, the relatives of the groom partake of the food that was placed on the five leaf-plates. The bride's mother-in-law now takes the couple to call upon the women-folk of the village landlord. Then, after dancing and singing on the part of the hired entertainers, the bride receives gifts and the party returns home.

[In other places, after the maṁdha has been disposed of, the women take the bride and groom, tie their clothes together, and go to worship at cross-roads or at some place on a public highway. The bride carries in her hand a lota of water with a twig of mango in it. Both she and the groom carry sticks. A fire-sacrifice is made and puris are offered and then eaten by those who make up the party. Then the bride and groom enter into a mock fight with their sticks.]

The midday meal follows. In this the whole biradri takes part. And in some places the rice remaining from the dinner is piled up on the floor and a piece of haldi is hidden in it. The bride is made to kick it over, and then the rice is distributed to the poor.

The clothes of the pair are again tied together, a drum is beaten by a sweeper, and, with a brass tray of cakes and batasas carried before them, they proceed to the boundary of the village, to the place where the village godlings are worshipped, and there make an offering before fire. They go and come, singing noisily. When they return, a stick of dhak, or of the nim tree, or of the cotton plant, is given to each of them, and they beat each other seven times.

A day or more after the arrival of the bridal-party at the groom's home a mock battle is fought. A person takes upon his head a brass tray in which gur is placed. A number are chosen to represent the bride and a similar choice is made for the groom. They now struggle for the possession of the brass tray. The groom's party finally wins. Thereupon the groom's father unties the fastenings of the maṁdha. Then the bride's mother sprinkles red powder over the groom's party. On this day (or a few days later) the bride's father and her brothers come to take her back to her own home. The negi (boy's maternal uncle) receives them and washes their feet, and they give four pice and sometimes make presents.

When the barat starts with the bride, powdered red peppers, mixed with urd flour, are rubbed over the faces of the groom's father and his relatives. It causes a good deal of sneezing. A shrub bearing burs (*māmrī*) is thrown over them and the burs stick to their clothes. The formal weeping on the part of the bride's female relatives follows; after which one of the women hides the bride in her lap, covering her with a sheet (chadar). Another bit of horseplay follows. The bride's mother makes puris and other delicacies, draws figures of horses and asses upon them, and gives them to the groom's father to take home to the groom's female relatives. She addresses the man thus: "Our horses and asses are taking by stealth our food for their women." This causes a good deal of laughter.

If the pair are of the proper age at the time of the śadi, the marriage is then consummated; otherwise, the bride, upon her return, remains with her parents until she is thirteen or fourteen years of age. Occasionally, after the parents first come to take her back home, she is again taken to her husband's home for a few days, and then brought back again by her father. In this case, also, she will then remain at her own home until she reaches the age of puberty. When the parties have reached the proper age for marital relations, that is, when the girl is about fourteen and the boy about sixteen, the boy's father sends word that he will come on a certain day for the girl. The date for this event, which is called the *gaunā*, is usually fixed by a Brahman not long after the wedding. The gauna must be in the first, second, fifth or seventh year after the wedding, but not in the third, fourth or sixth year. The day before the groom's party goes for the bride, the clansmen, or as many as wish to go, gather. The groom's father takes a chadar and a jacket for the bride, and two and a half seers of sugar and two and a half seers of rice for the bride's parents. They start out in the evening accompanied by dancers and music. When they arrive, the negi (on the bride's side) washes the feet of the groom and of his father and relatives. The gifts are then made over to the negi. Then coarse sugar and curds are served to five men. A feast follows, and liquor is drunk.

The next morning a wave ceremony may be performed. If so, the bride's veil is tied to the groom's clothes. Afterwards they are led to the conveyance in which they are to be taken to the husband's home. Before they start the bride's father may make other presents, such as clothes, jewellery and money. When they reach home, the negi's wife washes the bride's feet, and food is offered to her. At the door the bridegroom's sister demands a present, which must be given before they enter the house. The marriage is then consummated. After a day or two the bride may return to her mother's home. Before the *gaunā* the groom has no right to enter the girl's room or to touch her bed.

The rich may go again for the bride, following much the same ceremony as in the gauna. This is called the *raunā*.

During the wedding ceremonies a strict account is kept of the gifts received from the various relatives, so that proper return may be made later when other marriages occur. Part of the wedding expenses are met by subscriptions, which must be paid back double to those who gave them, when weddings occur in the donors' homes.

Analysis shows that there are three important divisions in the marriage. The first is the mamgni, which is to all intents and purposes binding. This is performed in infancy. The second is the śadi, or marriage proper, which usually takes place in childhood. And the third is the gauna, or the consummation of the marriage, which takes place when the parties reach the age of puberty. Authorities differ as to the indispensable parts of the ceremony. But all agree that the phera is essential to the wedding. Some add the sindurdan, or the marking of the parting of the bride's hair with red-lead. Among other essentials are sometimes named the kanyadan, or giving of the girl in marriage; the pamw puja; the eating together of the bride and groom before relatives, and the ceremony in the kohbar, or retiring-room. The essentials according to the *daiva* ritual of the Hindus are the worship of the gods, the fire-sacrifice, the gift of the daughter, and the phera.

A variation of the śadi ceremony is called the *dolā*.[1] This is a less respectable form of wedding used when the parents are poor. The chief point of difference from the śadi is that the barat is not received by the bride's parents. A Brahman is consulted as to an auspicious day for the ceremony. Upon that day the groom's father, accompanied by some near relatives, goes for the bride. They tarry near her home for refreshments and arrive at the bride's house at sunset. More liquor is procured, and a dinner is given by the groom's party. In the morning, after eating some parched gram, the party returns home taking the bride with them. Some of her old female relatives go with her. She is received by the boy's mother; the groom's younger sister worships her feet by pouring water over them, and then she is taken into the house.

After a few days, the wedding preparations are begun. All the preliminaries for both bride and groom are performed here in the groom's home. The preparations are

[1] See account in Crooke, *Tribes and Castes of the North-Western Provinces and Oudh*, Vol. II. p. 181.

as in the śadi proper. The lagan for the bride is followed by two sacred earth ceremonies (one for each), and then two marriage sheds are built. Then for both the anointing is performed, and for each a kohbar is prepared. The evening feast (bhaktawan) follows.

Male relatives of the bride come for the wedding ceremonies. On the day of the wedding, the groom is prepared for the barat, which simply marches through the village and returns. Then the bride is prepared to receive the groom's party as if she were in her own home. The groom's mother performs the wave ceremony and goes through the play at the well. Then follow the regular wedding ceremonies, including the foot-worship, the bridal bath, the fire-sacrifice, and the phera. The ceremonies are concluded with a wedding feast.

The next morning the services for the departure of the barat are carried out as in the regular śadi, but the barat simply marches around the village and returns to the groom's home. That evening they march around the village again, but before they start the bride's father performs a wave ceremony. The bride may go back home with her father. But she usually comes to her husband's home to stay.

A woman may not be married by the regular ceremony a second time. The rule is in agreement with Manu.[1] The marriage of a widow is therefore a formal acceptance of the woman by the man in the presence of witnesses, usually relatives. Such a union must be ratified by the panchayat. The groom may give a suitable dinner in honour of the event. The presents which the groom makes include articles which will remove the bride's signs of widowhood. Where the practice of the levirate enters in, the formal acceptance of the bride is called the *karāo*. The marriage of widows is also called sagai, but this seems to be a more general term than karao. Other terms are applied to widow-marriage, but these are more or less descriptions of the formal acceptance of the bride. Among these terms are *chādar*

[1] VIII 226; IX 47.

ḍālnā, *chādar uṛhānā*, and *dhaurānā*, and they refer to the covering of the bride with a chadar or sheet. Sometimes a woman answers the questions about having been married a second time by saying, "*Pañchāyat huī*," or "*Khānā huā*," or "*Rasm rasām huī*," or "*Daṁḍ huā*," which mean that the elders have agreed to the match, or that the feast or other formalities have been observed.

One form of the ceremony for the marriage of a widow with a widower is as follows: The relatives of both parties consult a Brahman as to an auspicious time for the event. At the proper time the groom, with a few near relatives, goes to the bride's home, taking two and a-half seers of sugar, two and a-half seers of rice, one loin-cloth, and one chadar, as a gift. A feast is then given by the bride's parents. After the feast, the groom withdraws with the bride to a private room. In the morning the groom takes her to his home. A bit of silver, upon which a figure representing the groom's deceased wife is engraved, is brought from the jewellers. *Dāl pūrīs* are prepared and served on five leaf-plates. These plates are arranged in a circle, in the centre of which a fire is lighted. The image and a new loin-cloth are placed before the fire. Then one plate is taken by the bride and the others by four women. The pair place red-lead on each other's forehead in the hair-parting. The bride then puts on the loin-cloth, and the image is hung from her neck by a red, or a black, string (or thrown into a well). The image is called the *saut sāl*. In the evening a few of her near relatives come to the groom's house and are given a special dinner. A short time after this wedding, the bride will be given ornaments besides those which she received before. This shows that the state of widowhood has been superseded.

In a much less elaborate form of the ceremony the groom, with a few of his friends, takes such clothes and ornaments as he can afford, and a box of vermilion, and goes to the house of the bride at nightfall. There the usual formalities are gone through, and, in the dark, small hours of the night, he clothes the bride, applies the vermilion to her forehead, and takes her home.

In another variation of the ceremony, the woman is dressed in new garments and presented by her husband with bracelets, a nose-ring and ear-rings, or some other emblems of wedded life. The man and woman are then seated together, and a white sheet is thrown over them (chadar dalna), in the presence of the brotherhood by an elder brother of the groom. Presents, or a rupee, are placed in the bride's hands. A feast follows.

In case the widow is married to a bachelor, the essentials of the phera are performed in the groom's house. But the groom does not perform the phera with the bride. In her stead, a piece of cotton-plant is tied to the plow-beam in the pavilion. The various parts of the śadi ceremony are performed for the groom, but not for the bride. She has already received these attentions at her previous marriage. It is this fact that accounts for the absence of the śadi ceremonies in other widow-marriages. In the above ceremony we have an instance of mock-marriage, *e.g.*, the groom is united with the cotton-plant. He afterwards receives the widow in marriage by the sagai rite as his wife.

CHAPTER V

DOMESTIC CUSTOMS: DEATH: MISCELLANEOUS

WHEN it is clear that a person is about to die, relatives ask him about the distribution of his property. Then, as the moment of dissolution approaches, he is placed upon the ground (usually). A batasa, or a *peṛā* (both, kinds of sweets), is dissolved in water, which is then given to the dying man to drink. Or, Ganges water, or water in which metal has been washed, is given. Occasionally, curds, or milk, are given. Those who are able, fee a Brahman to bring a cow, and the dying man is made to seize its tail and is thus helped over the river of death (*saṁkalp*). If a person who has not passed middle life should die on a bed, it is believed that he would become an evil spirit. On the other hand, some fear to place the dying man on the ground, lest he should discover their intention and curse them. The body is laid with the feet to the south, the direction in which it is believed his spirit will travel. (In some places it is laid upon *kāṁs* grass or wheat-straw.) After death, arrangements are made for the funeral. The news is sent to the relatives and friends, and cloth, coloured thread, betel leaves, sandalwood, ghi, and bamboos are brought from the bazaar. Then a stretcher (*tikhṭhī*, *arthī*) is made of the bamboos. The body is rubbed with gram flour, and then washed with cold water by the relatives or those in charge of the ceremony. The body is set up against the wall on a plank, and water, as it is poured over the body, collects in a hole which has been dug under the plank. If the deceased be a man, men perform this office; if a woman, women.

In some places the water for the washing is drawn from a well, with the left hand, and brought in an earthen vessel. Metal is placed in the mouth. If the corpse be that of a woman, it is anointed with ghi. Then scents are sprinkled over the body and it is covered with a rough sort of winding-sheet made from cloth brought from the bazaar. Each relative who is able to do so brings a winding-sheet, and all the sheets are wrapped around the body. At the place of cremation (or of burial) all but one of these winding-sheets are taken off. One is taken by the nearest relative, who lights the fire, and the rest given to faqīrs and to the Dom, or sweeper, who furnishes the fire. In some places all these sheets are taken by the menial who gives the fire. A tailor is now employed to make the clothes. These garments are of white for a man, of red for a woman. Over the body now, as it is laid on the bier, a red or a white cloth is thrown. The red cloth is used for a woman. In the Central Provinces the bier is painted white for a man, red for a woman. At the four corners of the bier flags are often fastened. In other places seven flags are used, one at the head and three on each side of the body, or seven *pān* leaves marked with red-lead are used. Flags are often of red cloth, or of gilt, according as the body is that of a woman or a man. In the case of children there may be no flags. The knife with which the string was cut in making the bier is carried with the body. If the frame is tied together with rope made of kams grass, the rope is made by rolling it with the left hand uppermost and by drawing the hand towards the body.

If a married man die, his widow removes the ornaments from her wrists and ankles. They are broken if of glass, but if of metal they are kept. Her hair is let down, and is thereafter unkempt. If the widow be a girl who has never lived with her husband, she will simply lay aside her jewels for a time.

If the person die late in the afternoon, or at night, so that it will be impossible to carry out the funeral ceremonies until the next day, the great-toes of the body are tied together, so that the body may not increase in length unduly. Sometimes the corpse is measured and a reed of the exact length placed beside the body. Unless these precautions are taken, the prince of demons (*Baitāl*, *Vetāl*) may get possession of the body and cause it to swell. Others say that the body must be watched lest an evil spirit (bhut) take possession of it and cause it to rise up, and that the watcher must not be left alone lest he be attacked. They say that such things actually happened in the olden times. A lamp is kept burning and in cold weather a fire also. There is great fear of the body at night. Cases have been reported in which the dead man, when being carried to the grave, or to the burning-place, at night, has seized one of the bearers by the neck.

The relatives and friends join in mourning. Some burn the palm of the right hand with a hot copper coin. Others do this after they have reached the place of burial or of cremation. Two, in some places four, balls of flour (barley, wheat, or urd), or of rice cooked in milk, are made, two of which contain copper coins. Of the two containing coins, one is placed on the right, the other on the left of the body.

If the person die during the *pāñchak* (the first five days of the new moon, or the first five days of the "dark" moon), four *piṇḍās* are made. These are placed on the bier, two on each side of the body. They represent the members of the family. With the body they make up the number five, indicating the five days, and thus removing, as they believe, the necessity of another death in the family during the year.

When the funeral procession is ready to start, the husband (or wife) of the deceased marks the forehead of the dead seven times with sindur. The husband makes the marks with his fingers. The widow marks the forehead, using the finger of the dead husband. This indicates the dissolution of the marriage relation, and corresponds to the seven rounds of the phera. The bier is first lifted by the relatives, both men and women, and afterwards carried by relatives and friends. The body is borne feet foremost, so that the ghost of the dead person may not be able to find the way back. With the procession fire from the house is carried in an earthen pot or on a piece of dried cow-dung. This is for protection from evil spirits and also to light their huqqas (pipes). As they go they throw barley, shells (*kauṛī*), and *tālmakhāna*[1] seeds in front of the body. Sometimes the women follow the bier. In case there is no man to serve as chief mourner, the widow goes to the place where the body is disposed of. The other women go but a little way and then return home. Each one, as she drops out of the procession, puts a bit of earth on the bier. Any other person leaving the procession does likewise. As they go they cry out, "*Rām, Rām, sat hai, sat bolo gat hai*" ("God is real (true); to speak the truth is salvation"). And they express their helplessness in such words as: "*Tū hi hai, taine paida kiyā, taine mār diyā*" ("Thou art God; thou hast created and thou hast destroyed"). The procession is sometimes accompanied by a low-caste man beating a *ḍhol*.

After the body has been taken from the house all the water-vessels are emptied, and such earthen utensils as the deceased had touched just before the time of death are broken. In some places water is sprinkled over the the bed to make it cool for the spirit of the dead. The dust of the room and the clothes of the dead are gathered up, carried by a relative and thrown outside the village. This applies to cases of infectious diseases only; in other cases clothes are given to the poor or used by relatives. In some places the spot where the body was prepared for burial is burned over with fire, or the place is liped. A fire is kept burning in the home for some time, usually three days.

On the way to the place of burning, or burial, the body is placed upon the ground five or seven times (*mansil denā*). At the first place where they stop barley grains, shells, and talmakhana seeds are left. In case the two pindas were placed on the bier, one, or both, are left there. It is from this point that the women usually return.

Chamars both bury and burn their dead, and there seems to be no fixed rule that determines the matter.

[1] Seed of the water-lily, considered as strength-giving.

Sometimes the poor, instead of burning the body, merely scorch it on the face and then cast it into some stream. The cost of wood is too great to admit of the poor burning their dead.

The usual custom seems to be to burn the body; and where there is no river near by, the ashes are collected, together with small pieces of bone, placed in a *ghaṛā* (an earthen pot), carried to some stream and cast into its current.

If the body is to be burned near a stream, it is placed upon the bank as soon as the procession arrives. It is then carried into the stream, sideways, and immersed. The bier is then placed on the bank, and the body is taken off and laid on the pyre, the feet towards the south or towards the stream. Wood is placed upon the corpse. Sometimes ghi is sprinkled over the body. In some places, a copper coin is placed in the right hand, and a live coal is placed upon the coin. Fire is brought from a Ḍom, or sweeper. The chief mourner takes the fire and walks round the body seven (or five) times, and at each round sets fire to the pyre at the head. If the head does not burst in the burning, it is broken with a bamboo from the bier. When the body is fully burned, the ashes and pieces of bone are washed into the stream.

If the cremation has taken place near a river, all remain to the end of the ceremonies. If the body is burned at a distance from a stream, no one is left to wash the ashes, but the mourners come the next day and collect the ashes and bones in a ghara and take them home. The earthen vessel is set aside, or buried, or hung on a tree, until someone takes it to the river Ganges. If the stream near which the cremation takes place is not the Ganges or one of its branches, or if the body was burned outside the village, by the village tank, the ashes are collected in an earthen vessel and afterwards taken to the Ganges.

In the Punjab, bones and ashes are watched to protect them from evil influences.

When the cremation is completed and the remains have been cast into the river, some member of the party says: "On this side and that side of the Ganges the son of So-and-so, the grandson of So-and-so, performs the pinda ceremony for his dead father; Mother Ganges accepts the service." The relatives then bathe. The knife which was put upon the bier is then set up in the stream and all present pour water over it, thus worshipping the Ganges and the fathers (*pitṛis*).

Members of the Śiv Nārāyaṇ sect bury their dead. While the preparations for the procession are in progress the *Santvirāsa* (Scriptures) are set up on an improvised platform covered with a white cloth. Before this, those present worship and then join in singing. Musical instruments are used. During the funeral procession chosen men (mahants), going before the body, read from the Santvirasa. As they proceed they sing. Some of the songs are songs of rejoicing. If anyone leaves the procession, he puts a lamp of earth on the bier.

At the grave the bier is placed upon the ground, then seven wicks are made of cloth, dipped in ghi, lighted, and waved around the head of the body seven times; and at each turn, the lips are scorched with the flames. Then a *laḍḍū* (a sweetmeat) is placed in the mouth. Camphor, or incense, or sandal-wood, is burned in the grave, in an earthen pot, before the body is let down. The corpse is taken from the bier and laid in the grave so that the feet are in a southerly direction. After the body has been lowered, earth is thrown into the grave. First the reader (mahant) casts in a handful of earth; then the chief mourner casts in five handfuls; after him all the others cast in earth. Then the grave is filled up. On the grave the flowers and the seven flags from the bier are placed. The grave is dug in the usual way and then a chamber is hollowed out on the side large enough to receive the body. After the body is laid in this side-chamber it is walled up before the grave is filled. In this way the earth does not fall directly on the body. Some members of the sect burn their dead, and still others cast the body into a sacred stream. Children and persons not initiated are buried without any ceremony. If the wife of an initiate die, her relatives may claim the body and burn it.[1]

Dādū Panthīs[2] burn their dead at dawn; but the more religious not infrequently request that their bodies, after death, be thrown into some field or some wilderness, to be devoured by the beasts or birds of prey, as they say that in a funeral pyre insect life is apt to be destroyed.

The funeral ceremonies of the Kabīr Panthīs are described as follows[3]: "Upon the death of a member of the *Panth* two cocoanuts are immediately purchased. One of these is carried by the barber in the funeral procession and placed by the side of the dead body immediately before cremation or burial; the other is kept in the house and reserved as an offering at the funeral *chauka* to be held at some subsequent date.

"The arrangements in connection with a funeral (*chauka*) differ from those of an ordinary *chauka* in that the awning over the prepared ground is of red instead of white material; a piece of white cloth is placed over the *chauka* to represent the dead man's body, and the number of betel-leaves is reduced to 124, the leaf removed representing the dead man's portion.

"At the commencement of the service the mahant prays silently on behalf of the deceased, that he may be preserved from all dangers on his journey. Upon the conclusion of this prayer five funeral *bhajans* are sung, after which all present do *bandāgi* to the guru three times and to the piece of white cloth that represents the body of the deceased.

"The cocoanut which has been specially reserved for this service is next washed by the mahant and made over to some relative of the deceased, or, should there be no relative belonging to the Panth, to some member attached to the same guru as the deceased. This man, after applying the cocoanut to his forehead, shoulders, etc., returns it with an offering to the mahant, who breaks it upon a stone upon

[1] See Crooke, *Tribes and Castes of the North-Western Provinces and Oudh*, Vol. II. p. 188.
[2] *The Religious Sects of the Hindus* (C.L.S.), p. 54.
[3] Westcott, *Kabir and the Kabir Panth*, pp. 133, 134.

which camphor is burning. The rest of the service is conducted in the manner already described. The number of cocoanuts offered varies from one to nine, according to the means of the friends and relatives. Each cocoanut involves a separate offering to the mahant. The flesh of the cocoanut or cocoanuts is made up, with flour, etc., into small cakes, which are sent round to the houses of Kabīr Panthis, by the hands of Bairāgīs."

Satnamis burn their dead, laying the body with the face downward and spreading clothes in the grave, above and below the body, to keep it warm and comfortable. Mourning is concluded on the third day, when the relatives have their heads shaved (but not the upper lip).

Other sects, *e.g.* the Rai Dasis, the Nānak Panthis, and some other Rāmats, bury their dead, unless the request has been made during life that the body be burned or exposed.[1] Sometimes members of sects following Rāmānanda read from the Rāmāyaṇa, at the burning ghat, to help the deceased on his journey in the spirit world.

When a person dies of smallpox, plague or cholera, the body is disposed of as soon as possible. Sometimes it is buried. Usually the body is not burned, but cast into a stream without any ceremony. But, on the day of the death-feast, an image (about twelve inches long) of the man is made of flour. Over this the ceremonies of burial or cremation are performed.

If a person die away from home, the body is burned or buried immediately. The relatives, when they learn of the decease, make an image of the dead person and perform the funeral rites over that.

At the place where the body has been disposed of, after the knife is worshipped, a small fire is lighted and ghi and parched rice are offered in it. They then say, "In whatever state you are, leave us alone." This expresses the feelings of the people and the aims of the ceremonies, and especially of the customs about to be described. Another formula is used, whenever food is offered to the departed:

[1] Crooke, *Tribes and Castes of the North-Western Provinces and Oudh*, Vol. II. p. 183.

"Be kind to us all; make us prosper who are left behind." The desire is to prevent the return of the spirit to its former home. They take precautions, however, so that, should it return, it may be propitiated and do no harm.

When the funeral ceremonies have been completed, the company bathe and start home. As they leave the place they throw earth backwards with their left hands. They do not look back. On the way back one or two cakes of gur and parched rice are distributed. Some plant a few stalks of grass near a tank as an abode for the spirit, which wanders about until the death ceremonies are completed. Water is poured here for ten days.[1]

Before returning the party also partakes of sweets. Up to the ninth or tenth day, the chief mourner carries a lota and a knife; smokes by himself; cooks by himself, not using a tawa, and does not eat salt. He has no intercourse with his wife, and sleeps on the ground. A piece of the winding-sheet is worn by the chief mourner, about the forehead or neck, during the days of mourning. When a *very* old man dies his sons each wear a piece of the winding-sheet for a long time.

In many cases no precautions are taken to bar the ghost, unless a number of the members of the family have died in close succession. In such cases sarsoṁ-seeds are dropped on the way as the funeral procession goes to the burial-place or the burning-gounds.

While the burial, or cremation, ceremonies are in progress, the women who have gone back home prepare for the return of the procession. They bathe, and make sherbet to refresh the party upon its return. In the larger towns the men stop at the liquor-shops on the way home.

At the door of the courtyard each person, upon his return, touches water, sometimes in a broken earthen vessel, fire, a sil, and iron (*chimtā*, fire-tongs) and nīm-leaves with the great-toe of his right foot, and takes up a nīm-leaf and bites it into two pieces with his teeth. This

[1] Crooke, *Tribes and Castes of the North-Western Provinces and Oudh*, Vol. II, p. 184.

last act signifies that the relations with the dead are dissolved. They then enter the courtyard (or house), where they partake of sherbet, or sparingly of gur. Then the relatives depart. If a young person die, no food is eaten on the day of the burial (or cremation). If an old person die, food is prepared by relatives and brought in. Of this food each offers a small portion of the first roti in the fire on the spot where the body lay. The remainder of this cake is thrown out for the dogs.

The next morning rice and dal with milk are prepared for food in a new earthen pot. Some of this is put upon a leaf-platter and placed outside for the dead, and water is poured out nearby five times. This is for the spirit of the dead. After that the family partake of the food, but not of the rice-water. The chief mourner eats first. He may eat salt in this meal. A variation of the practice is that the chief mourner places outside of the house of the deceased an earthen pot full of milk and rice-gruel, with a pitcher of water, for the departed spirit.

On this day kus-grass is planted where the chief mourner will bathe during the days of mourning, and he pours water on this with his hands each day for ten days. He eats once a day.

There is some variation in practice concerning the feasts and ceremonies. In some places certain matters are attended to on the third, while with others these are left until the tenth day. Again, the principal feast is sometimes held on the third instead of the tenth day. During the first three days after the funeral those who carried the body sleep on the ground. On the third day (*tri rātri, tījā*), in some places on the tenth day, oblations and cakes of barley flour are offered to the departed soul. A mesh-bag is hung up in the door, and in this a dish of water and a little food, when ready, is placed for the spirit. The relatives cook rice and urd for this tija feast, and it is eaten in the house where the dead man used to live. Four cakes from the meal are set out upon the roof on plates of dhak-leaves. In some places a trench is dug on the right side of the door, and seven pots of different kinds of food are buried in it, and milk, mixed with Ganges water, is taken to the place of cremation and sprinkled about. On the fire that was lighted outside the house, where the body was placed, a small dish of rice is cooked. Portions of rice are placed on leaves (of *pipal* or of banyan) and with them each man who helped to carry the body has his shoulder touched. These leaves, the rice remaining, and the vessel in which it was washed, are then burned on the spot. Sometimes the men who carried the bier have cross-marks made on their shoulders with sarsoṁ oil. In some places these four men wave a little of the rice over their shoulder with their left hands. In this feast the four persons (*kandhaurī*) who were the first to carry the dead (they are relatives) are especially feasted; and after this they are absolved from contamination and may resume their beds. On this night, ashes are sprinkled in the doorway (outside and inside). Often the ashes are covered with a basket slightly raised from the ground on one side with a stick. In the morning search is made for the footprint of some creature, for this footprint in the ashes will indicate the nature of the spirit's new body.

Chamars believe that the dead return now and then, and provision is made for the visits of spirits. On the third day, especially, does the spirit return; hence hearth-ashes are scattered at the door at this time.

For ten days after death, food for the refreshment of the spirit is placed outside the door, on the roof, or under the eaves of the house, and some is left on the way to the cremation-grounds and at the burning-ghat or at cross-roads. A thick loaf of bread is sometimes given to a cow. The earthen pots of the house are broken that the spirit may have water-vessels. Some plant a few stalks of grass near a tank as an abode for the spirit, which wanders about until the funeral rites are completed. On these blades water is poured daily for ten (or thirteen) days.

When these ceremonies are performed on the third day, they are repeated on the tenth day, and the clansmen are fed. A portion of the food is placed on the road where the first cake (pinda) was left on the day of the funeral.

On the tenth day the chief mourner, or the nearest male relative, is shaved. Other near relatives have their hair trimmed. The chief mourner changes his old clothes for new ones, and gives the cast-off ones to the barber. A small platform is made on the bank of a tank or a stream and covered with sacred grass. On this balls (pinda) of *arwah* rice, cooked in a new pot, are offered by the chief mourner, one ball being offered for each person of the family who has died; that is, balls are offered for brothers, parents, grandparents, and great-grandparents. Then, taking up the balls, he touches his right shoulder then his left shoulder with them (the reverse order is followed if the chief mourner be a woman), and places them on the ground at some distance from the spot, or casts them into a stream. The rice for the pinda is cooked near a stream or tank. No chulha is used. All now return to the house and partake of the feast already prepared. This is made of arwah rice and dal cooked in a new vessel. Part of this is set out for the dead. Then each relative receives a little of this food. The meal is served on leaf-plates. The chief mourner receives the first helping; and, after food has been placed before all those present, a little from each leaf-plate is taken, placed upon a leaf-platter and set outside, by the chief mourner, for the deceased. When the chief mourner returns, he eats five small portions of the food and then gets up and washes his hands, as though he had finished the meal. He then greets the company, saying, "*Lakshmī Nārāyaṇ, pañcho!*" Then all begin to eat. He sits down and joins with them. Liquor is drunk. These ceremonies are a purificatory rite. In case the chief mourner be a woman, these rites are performed on the ninth day.

On the evening of the tenth day another feast is prepared. After the meal is spread, before anyone eats, a portion from each plate is placed in a leaf-platter, taken outside and left for the deceased. Then the chief mourner begins to eat. After taking five morsels he washes his hands and greets the company as he did in the morning. Then all join in the feast. Liquor is provided, and some is drunk before the meal. If the deceased was an old person, singing and dancing by boys of the caste is provided. These professional dancers are paid for their services. This feast lasts late into the night. The women and children eat after the men have finished. At the close of the meal a basket is taken and into it are put the new earthen pot used in cooking the meal on the morning of the first day, that used in preparing the meal on the morning of the tenth day (or ninth), and, if the deceased was a woman, a small pot containing oil, and, if she was in the habit of smoking, a mud huqqa with a chilam, in which is tobacco and fire (ready to be smoked), and a broom made of a special kind of grass. If the deceased was a man, the broom is not used, but a cover is placed over one of the vessels so as to make up five articles in the basket. If the deceased was not a smoker, a small vessel takes the place of the huqqa. The articles in the basket must number at least five. Five men join in the procession; one, the sister's husband, or sister's husband's father, carrying the basket, and followed by one carrying fire, one carrying fire-tongs, and one carrying a knife. These articles are provided for the use of the spirit of the deceased. The last man carries nothing. The basket is deposited outside, and the men, bringing with them the tongs and the knife, return. Then a little food is given to each of the five from one plate.

After this feast, when the guests have left and the people of the house have gone into another room, a widow of the family takes two new earthen plates, on one of which she places urd ki dal and on the other *chana*, or some other kind of dal. Then she sifts ashes from the hearth-fire over a small space in the room and covers them with the sieve. The two plates are placed near the ashes. She sleeps in the room. In the morning the ashes are examined for footprints. If no mark is found, the conclusion is drawn that, for the deceased, the round of transmigration is finished, or that the spirit has been "laid." If someone in the family falls ill soon afterwards, a bhagat is called, who may report that the ghost has become a wandering evil-spirit. The woman who slept in the house is given a dhoti or a rupee as a reward,

Some make a mark on the body of the deceased with ghi, oil or soot; and when a child is born in the family, its body is examined, and if a corresponding mark is discovered, it would indicate that the spirit had taken its new birth in the family. Some make a test with ashes at the annual funeral-feast and at Dewali-time, to discover whether the dead has paid his former home a visit.

In some parts of the country, on the eleventh day, or on the night of the tenth, the utensils and private property of the deceased are made over to his sister's husband; but in other places he receives a lota, a brass tray, or a rupee. The feast of the tenth day, which is the principal death-feast, is called *Dasa Pitar* and *Visarjan.*

A tribal feast is sometimes given on the twelfth, thirteenth or sixteenth day after the funeral. The relatives come to offer consolation, and they must receive refreshments. There are places where the principal feast is on the thirteenth day. It is cooked on a special place plastered with dung from a cow that has not calved. The food consists of rice and *shakkī.* Then sindur and food are offered to *Bhūmiā* and other-godlings. Up to the time of this feast the chief mourner is under certain tabus; for he has worn only scant clothes, and a handkerchief on his head, and has carried a lota with him; has not worked, has not slept in bed, and has made offerings of food for the dead. The house is liped, and rice cooked in milk is served. Offerings are made to godlings and to the dead man. Portions of the food are provided for Brahmans and neighbours and served on small dishes of leaves. Likewise food is set out on the roof for the crows. A fire-sacrifice is made, with offerings of ghi and halwa, to the dead. After the offering the feast is spread, and drinking is indulged in. The dinner takes place in the night, and then the guests depart.

Food is given to the sweeper and to the dhobi and to other menials after the feasts.

In some places, after a month and a half, that is after three half-moons, a feast is held in the name of the dead. No special kind of food is prescribed. The relatives assemble at night. The offerings for the dead are made as on the other feast-days. When the meal is served, a little food is taken from each plate and carried out on a leaf-platter by the chief mourner and left for the dead. The chief mourner eats five morsels and performs his ablution as at other death-feasts, and then they all join in the meal. There is drinking also. In the first half of the month of *Kātik* (*Pitar-pakh, Pitar-paksh*) the bones of the dead (*phūl*), if burned at some other place, are carried to the Ganges (or a tributary stream). The chief mourner who brings them bathes, and then, holding sacred grass in his hands, pours water into the stream in the name of the dead, five times for each ancestral spirit (for three generations back) and for his deceased brothers. If the ceremony is performed with water from a well, he pours out water but once for each spirit. The offering of water is made each day of the first half of this month. A space in front of the door of the house is plastered with cow-dung and on this flowers are offered and flour is sprinkled. This is done for fourteen days. A feast for the dead is given, on the ninth for a woman, and on the fourteenth for a man, and again on the fourteenth for a woman. The last day's ceremony is for all the dead. After this no feast will be given for her on the ninth. At this time the feasts are held in each house, but anyone may have guests. On the eleventh day of Katik sixteen or seventeen balls of barley-flour are made. One of them is taken out and set aside for the Dom. Then, upon a platform made of clay and plastered with cow-dung, and over which sacred grass and leaves have been spread, a fire is placed. In the fire ghi and gur are offered. Then, in the name of each deceased ancestor, a ball is placed on the grass. If the balls are insufficient in number, the last is offered for all whose names have not been called. If there are too many balls, those left over are given to the ancestors collectively. After all the balls have been offered they are lifted up and raised to the right shoulder and then to the left, and then cast into the stream (or the tank if the ceremony is held there). On the last day the ceremony is performed in the same way. The chief mourner, or the one who performs these offices, is shaved. Some say that, until the Pitar-paksh is over, no wedding can be performed in the family. Others say that under a year no wedding can take place. Some hold that after the Pitar-paksh the gauna may be celebrated.

On the anniversary (*barsī*) of the death twelve pindas are offered, and the family, if they can afford it, give a dinner, and offerings are made to the dead as at other feasts. This may be repeated year by year.

The son, and probably the grandson, will keep up the offerings to the deceased. Brahmans are sometimes employed to make the offerings to the dead, especially those of the Pitar-paksh. In any case they receive gifts.

References have been made to means used to help the spirit of the deceased in its progress towards a peaceful reincarnation, and notices have been taken of acts which provide protection for those who are responsible for the funeral. Other references to means used to "bar" and to "lay" the spirit of the dead will be found in the next chapter.

In connection with the preceding ritual mention has been made of ancestor-worship. The whole of spirit-worship, both of the sainted and of the malevolent dead, so far as it deals with the ghosts of deceased relatives, may be considered as a form of ancestor-worship. The sainted dead are household guardians. Deified persons, like Nona Chamari, are considered as the ancestors of tribes or of sub-castes. A very large share of the attention given to the spirits of the dead is related to demonology and to magic in general; and this phase of the subject will find ample illustration in the next chapter.

That which more strictly may be called ancestor-worship occurs in the domestic ritual. When a son is born, and sometimes at the birth of a daughter, spirit is taken into the hand and waved about, and as drops of the liquor fall upon the ground the names of ancestors are called. At marriages some offerings are made to the spirits of the dead. But it is in the funeral rites that the greatest emphasis is laid upon the worship of the fathers. The effort made to supply the needs of the deceased are evident in the offerings of food, water, and utensils,

During certain festivals such as Dewali, and in the ceremonies of the Pitar-paksh, preparations are made for the return of ancestral spirits; for spirits continue to be interested in the affairs of the living. There is a social element in the funeral ceremonies, in the annual feasts for the dead, and in some of the Dewali ceremonies. There are also elements of fear in the intercourse with the dead. Freed from the limitations of the body, spirits move in a wider sphere and exercise greater powers. They can either harm or help, and one is never sure just which they will do. Some who have been elevated to sainthood are supposed to afford protection against certain demons and godlings.

In the north-west of these Provinces and in the Punjab ancestral shrines are found in the fields. The small ones are for ancestors and the larger ones for clans. Some of these are places of pilgrimage. Here and there the *Satī* is taking the place of these shrines. Occasionally images of the Sati and her husband are found. Her sacrifice has secured for her deification, so she is able to protect her worshippers and grant them their desires. Therefore women resort to the Sati, asking for children and other boons, and at marriages offerings of milk, food, fruit, and flowers are made to her. Neglect of the Sati may result in barrenness, or in disaster. In the east an earthen pot (*karwā*) with seven holes in it is offered. Other offerings consist of lights and food. Some put a lump of clay in the cooking-room to represent the ancestors, and an image of a ghost which makes trouble is set up in the house.

Regular worship of ancestors, conducted by the oldest son living, is performed by the offering of a goat. Sometimes this is performed in the ancestral shrine. And oblations and offerings are made on liped spots facing the south and in dreary and solitary places and on the banks of rivers. Daily oblations of water are poured out; grains of sesame and barley are used.

The house-worship is very simple. There seems to be no practice of bringing home the nuptial fire, or of keeping a sacred fire in the house. In some places a house godling is supposed to occupy a special mound on the floor, or in the courtyard, or a place in the wall, or in the thatch, or on the grain-bin. Here, in the godling's station, on the day of the Dasehra festival, seven wheaten cakes and some halwa are offered, and water, or water mixed with ground cloves and cardamoms, is poured out as an oblation. Sometimes the offering consists of a young pig and some spirits. Sitala often has a special place in the house. There is an element of house worship in some of the great festivals, such as the *Nāgpañchamī* and the Dewali.

In house-building, a Brahman is first consulted as to when the digging for the foundation should be begun; in the name of which man of the family the digging should be begun; where the door is to be set; and whether an evil spirit inhabits the spot. Then the Brahman indicates the direction in which the man who begins the digging should face.

If, in the digging for the foundations, human bones, or a considerable amount of charcoal, should be dug up, the site would be considered inauspicious.

When the laying of the foundations begins, sweets are distributed. Shells and pice are buried in the foundations. If, during the process of building, the walls repeatedly fall, a Brahman is consulted, because the trouble is attributed to evil influences. He announces the necessary offerings to be made to satisfy the spirit responsible for the trouble. Then follows the sacrifice of a cock, or a pig, or a goat, or a buffalo. Sometimes a human being is named by the Brahman. In that case, a person is sent up on the walls, on some pretext or other, and an "accident" is brought about, and his body is left in the foundation. This result is accomplished by stealth. This very rare practice is a relic of the older custom of human-sacrifice in connection with house-building. There are many superstitions about this practice in connection with large enterprises, both Government and private. This form of superstition is still common.

During the time the building is in process of erection, a lamp is kept burning at night, and droppings of pigs, or other filth is left around, lest spirits take possession of the building. An old shoe is tied to a bamboo, which is set up to ward off the evil eye during the process of erection. Sometimes an iron pot or an earthen pot painted black is set up to ward off the evil eye. When the door-casing is put in, a member of the family drives a nail into the door and on this hangs a mud bowl with a small neck (kulhiya). An iron ring is attached to this.

If the house have a courtyard, the door in the wall should not face the south, as this is the direction of the abode of the god of the dead. In general, houses should not face the south, nor should the fireplace. Likewise a man should not set his bed so that he must sleep with his feet to the south, unless he is about to die. If the house is set on the north side of the street, the door is often built into a little inset at an angle to the compass, or to face in the direction east or west. Another taboo relates to the shape of the courtyard. It is always nearly or quite square, the feeling being that a long narrow court in their houses is unlucky. A narrow courtyard resembles the Ganges, and suggests the possibility of the whole house or courtyard being carried away as with a flood. Or, a narrow courtyard resembles a snake.

When the house is finished, a Brahman is asked to fix the date and hour when the family may take possession. If the date is some days off and the house is urgently needed, another entrance is made. This is done with the consent of the Brahman. The dedicatory ceremonies are performed on the day when they enter the house. The chief godling of the village is worshipped. Ganesh is not worshipped. Then the family enter the house. Sometimes the wife's chadar is tied to the husband's clothes. Then, as the Brahman suggests, they sacrifice a cock, or a goat, or a pig, or a buffalo to their special goddess. This is done in the courtyard, or in front of the door. The blood is covered with earth. The flesh is served in the feast connected with the ceremonies. A fire-sacrifice is performed before or after the sacrifice, as the particular godling prefers, and in the house or outside as the Brahman may suggest.

Chamars have their part in the festivals of the land, and no special notice need be taken of these occasions as such. But there are domestic aspects of great festivals which may be noted.

The Holi is a spring festival, in which the firstfruits of the spring-harvest are offered. Characteristics of the celebration are the doll-swinging and the scattering of red powder, red liquid and mud. The Holi fire is lighted at night, or in the small hours of the morning. Fire from this bonfire is taken into the house for the women. From this a fire is made, upon which small cakes are cooked, of which each member of the house partakes. Sometimes a stalk of the cotton-plant is set up in the house-fire, and, when it burns and falls, the folks determine whether good or bad luck will follow the household during the year. If the stalk fall towards the east or towards the west, it is taken as a sign of good luck; if it fall towards the north or south, misfortune will be looked for. In the Holi fire handfuls of grain, in the stalk, are parched, and this is laid up in the house, and parched grain is put away in the roof, or in the grain-bin. The day on which the fire is lighted is given up to feasting, and it is a bad omen if one does not have a hearty meal. The Holi is, for the Chamar, a time when he is utterly abandoned to debauchery. On the night when the bonfire is lighted he gives himself up to drunkenness, excess, and obscenity. By the time that the fire is lit, he is completely under the influence of liquor. As he dances around the fire he breaks out anew into drunken and lewd revelry. The women sit in the shadows near their homes, and listen to the singing and to the utterly filthy jests and songs of the men. The debauch connected with the Holi is prolonged for several days. Then the women take sticks and go about the village, or town, demanding gifts of coarse sugar from all sorts of people. Their conduct is very unseemly. The Chamar seems to yield to utterly degrading elements in this new year's festival.

Another festival that may be studied in its domestic aspects is the Nagpanchami. This is held in the middle of the rainy season, in honour of snakes. The women plaster the house, or at least the walls on each side of the door, with clay or cow-dung. Then they bathe. Afterwards, at the threshold of the front-door, they make images of snakes out of cow-dung, and draw on the walls on both sides of the door, with lime or with cow-dung, lines to represent snakes. Sometimes a wisp of grain, tied in the form of a snake, is dipped in a fermented mixture made of wheat, grain and pulse steeped in water; and this, together with money and sweets, is offered to the serpents. Saucers of milk are set outside the house as offerings to snakes, and the worshippers join their hands in the attitude of adoration. Milk and dried rice are poured into the family snake-hole. Songs are sung in honour of the serpents. A line is then drawn around the house—this is a magic circle across which a snake will not pass. A fire is lighted and ghi is offered in it. A feast with carousing follows. It is a day of hilarity, and cattle get a holiday, special food, and an extra allowance of salt. After the day's activities the images are thrown away.

One more festival, the Dewali, may be mentioned because of its domestic aspects. At this time the houses are cleansed and freshly plastered with cow-dung or clay; old lamps are thrown out and new ones are brought in. This is the time when the ancestral spirits visit their old homes. The family light lamps and sit up all night to receive the family ghosts. In the morning the wife takes all the sweepings and old clothes of the house in a dust-pan and throws them out on the dunghill, saying, "May thriftlessness and poverty be far from us."

Meanwhile the *Gobardhan Dewāli* is performed by the women. It is made in honour of Krishṇa, and consists of a prostrate figure to represent him, made of cow-dung, surrounded by little mounds of cow-dung representing mountains. Stalks of grains tipped with bits of cotton are set up in the mounds to represent trees. On the "mountains" tiny balls are made to represent cattle, and other balls trimmed with bits of rag to represent men. On this Gobardhan the churn-staff, five whole sugar-canes, some parched rice, and a lamp are placed. The cowherds are called in to worship and are then feasted with rice and gur. Gambling and intemperance are the prominent elements in this festival, and men go beyond all bounds in indulging in these vices. This is also the time when the goddess of good fortune visits the homes of the people, and they prepare their houses for her visit. The Chamars take their shoemakers' tools, or other implements with which they earn their living, to the headman of the local village group, and at his house perform a fire-sacrifice before them. In some places the implements are worshipped during the Durga Puja.

This is a time when a good deal of magic is practised. One instance will suffice. A hoot owl, which has been carefully kept for a year, is furnished with an image of a tiger, upon which to ride, and is made drunk with liquor. If a man takes the kajal, made from the ashes secured by burning this owl's eyes, and rubs it into his eyes, he obtains magical power which puts under his control any woman upon whom he looks. On the other hand, a man who eats the flesh and liver of such an owl becomes the slave of the woman from whom he receives it. (Of course, this food is given by stealth.)

CHAPTER VI

THE SPIRIT WORLD[1]

THE Chamar is saturated with animistic ideas. For him, inanimate objects, trees, plants, animals, and even human beings, are the abodes of spirits. The phenomena of nature are a mystery explainable on the ground of the spirit world. Furthermore, the experiences of life are referred to invisible spirit-forces. To rude men the ups and downs of life seem to be dependent upon the mere caprice of this invisible host, and this shadowy company of unknown powers is responsible for calamity, fever, cholera, smallpox, and other untoward events. These fickle, treacherous inhabitants of the unseen world, the demons and the godlings of disease, must be conciliated: and the tutelary godlings, the sainted dead, and other well-disposed spirits must be enlisted against the forces of calamity and disease. The superstitious man, of necessity, is always on the alert to outwit evil and malignant spirits and to circumvent their undertakings.[2]

The worship of stones is universal. The respect which the Chamar pays to them is independent of the shape or finish which they may possess. Village godlings are represented by stones, and occasionally stones well-carved are found in the house and at the village shrine. As a usual thing the stones representing the village godlings are

[1] On the various topics in this and in the following chapters see Crooke, *An Introduction to the Popular Religion and Folklore of Northern India*.

[2] See *Imperial Gazetteer*, Vol. I. pp. 473, 431; *Census of India*, 1901, Vol. I. Pt. I. p. 352; and Whitehead, *The Village Gods of South India*, p. 145.

smeared with vermilion, a survival perhaps of the ancient blood-sacrifice. This collection of stones under the tree on the village boundary is one of the few groups of godlings to whom the Chamar has access. Stones play a part in the cure of disease. The stone-mill and the sil and batta are fetishes.

It is easy for simple folk to believe that spirits live in trees. Motion is a sign of life; and, besides, the winds, passing through the trees, produce sounds which are heard as voices. Trees should not be disturbed after sunset. People are loath to cut down living trees. In cleared lands some trees are left standing, especially those which are known to be inhabited by spirits. The planting of trees, on the other hand, is a meritorious act, and it is often done with the hope of securing offspring, or increase in cattle. There are many trees held in special veneration. This is illustrated in their use in the domestic ceremonies, in the practice of magic, and in the exorcising of disease.

One of the most widely venerated trees is the pīpal (*ficus religiosa*), (and its near relatives, *e.g.*, the banyan). The worship of this tree, which may be of totemistic origin, is connected with the care of the dead and with the desire for children. Every leaf of the tree is said to be the abode of a god.

The nīm tree (*azidirachta indica*) enters very largely into the Chamar's superstitions, and is perhaps more universally revered than any other. In some instances its worship is of totemistic origin. Its leaves and branches are used in various phases of the practice of magic and in the barring of ghosts; and it is the abiding-place of Sitala Mata, the goddess of smallpox. With it are connected sun and snake worship. Fresh leaves of this tree are applied to snake-bite wounds, and sometimes given to the sufferer to chew.[1] Its leaves are used in many ceremonies.

The mango enters largely into superstitious usages. Its wood and leaves are connected with the practice of magic, especially that relating to fertility, and its wood is used in sun-worship and in the fire-sacrifice.

[1] If they taste sweet, he will die; if bitter, he will recover.

The mahuā (*bassia latifolia*) and the babūl (*acacia arabica*) are of great economic value. Besides this, the former is inhabited by spirits, and the latter is used in witchcraft. By pouring water upon a babul tree for thirteen days, a person will obtain possession of the spirit of the tree. It is believed that a person sleeping on a bed, the legs of which are made of babul wood, will have bad dreams; and that the ghost of a man burnt with this wood will not rest quietly.

The bel (*aegle marmelos*) and the ḍhāk (*palasa*) are venerated also. The latter is used in the marriage ritual; and from its flowers the red powder used in the Holi is made. The wood of the dhak is used in the fire-sacrifice. Both bel and dhak leaves have medicinal qualities.

The gūlar (*ficus glomerata*) is useful in the practice of magic, as is also khair (*acacia catechu*). The latter protects one against magic spells and the evil eye, and wizards keep away from its shade.

Besides these, there are various kinds of trees, such as the semal (*bombax hephtaphyllum*), the siris (*acacia sirsa*), the sāl (*shorea robusta*), and the jhund (*prosopis spicigera*), whose worship is more or less of a local character.

There are many trees which are pointed out as the abodes of particular spirits. The *Churel* lives in a broken tree, or in a tree in the jungle; and the terrible *Dano* and the giant demons (rakshas) have their special tree abodes. It is dangerous to go near these trees, especially late in the night.

The bamboo, the cocoanut, and the plantain are used in ceremonies related to fertility.

The leaves of the tulsi (holy basil) are used in worship and as a medicine.

The serpent is feared and worshipped. Offerings of milk and rice are made to secure the goodwill of snakes, and they are addressed with euphemistic titles to secure immunity from snake-bite. The first milk is sometimes offered to *Nāg*. The worship of *Rāja Bāsuk*, or *Vāsuk*, the chief of serpents, is famous; and the legends of *Gūgā* or *Zahrā Pīr* deal with the control of snakes and protection from snake-bite. The great cobra (*Sīs Nāg*), in shaking his head, causes the earth to quake. The snake is the emblem of longevity, since it renews its life from time to time, and it is sometimes looked upon as an ancestral ghost. The black snake (cobra) is the guardian of cattle and of water-springs. It is believed that snakes can prophesy; that they can spit fire; that they can burn anything with their breath; and that they guard hidden treasure. There is a widespread belief in the snake-jewel, a stone, or a silky filament which is spun by, or spat out by, a snake a thousand years old, on a dark night, when it wishes to see. This jewel is luminous. To obtain it a person must throw a bit of cow-dung upon it. The jewel is very valuable, since it gives immunity from all misfortune and the realization of every wish; and since it also preserves from drowning. This jewel is an antidote to snake poison.[1] Chamars kill snakes. There is also a belief in dragons, and certain caves, like that of Kausambhi, near Allahabad, are named after such creatures.

Various animals are venerated. The horse, while not actually worshipped, is considered a lucky animal. It is believed that the marks on his legs prove that he once had wings. His images are used in Guga worship; and at the shrines and platforms of certain saints and godlings images of horses are found.

The donkey is sacred to Sitala. The belief that he sees the devil when he brays is of Mohammedan origin.

The dog is the vehicle of Bhairoṁ, and he is connected also with beliefs concerning the god of the dead. The black dog is worshipped as a *Jinn*, and its grave is sometimes honoured. Its secretions are used to scare demons. It is fed to save children from dog-bite, and from other diseases and sicknesses.

The cat is an object of reverence. No woman will strike a cat, because it is the vehicle of Sati. If a person kill a cat, he must beg for a time, and then go to the Ganges and bathe. Afterwards he must sprinkle Ganges water upon his food, and give a feast to his neighbours.

[1] A similar jewel, or stone, is found in the forehead of the frog that jumps and catches birds.

Chamars believe that a cat has power to make a person temporarily blind. This she does in order to steal his food. An instance of magic is found in the belief that the after-birth of a cat rubbed on the eyes enables one to see in the dark.

The goat is worshipped, the black goat especially being prized for sacrifice. The goat is used also in divination.

Both the cow and the bull are considered sacred. The five products of the cow are very efficient scarers of demons. A cow helps the departing spirit over the river of death. The Chamars bow before the cow. In some parts they will not eat beef, although they will eat of the carcass of a cow that has died. The male buffalo is sacred to Kali. At the time of purchase Chamars worship both buffaloes and cattle.

The black buck and the elephant are also worshipped.

The monkey is worshipped in connection with the cure of barrenness; and, as Hanuman, has become a tutelary godling in every village.

The tiger (and the leopard and the panther likewise) is worshipped, and parts of its body are used in various ways. Tiger's fat cures rheumatism; its heart and flesh are tonics; and its flesh is burned in the cattle-stall to dispel cattle disease, and in the field to ward off blight; and its whiskers and claws are of great value as charms. Witches can turn themselves into tigers, and men are sometimes so transformed. A tiger without a tail is thus explained. A man-eating tiger obtains possession of the soul of the person whom he eats. The tiger has titles of divinity, as, Bāghadeo and Bāgheswar.

The alligator (*magar*) and the crocodile (*ghaṛiyāl*) are held in respect and their flesh is valued.

Jackal's flesh is used in the practice of magic.

Many birds are respected. The pigeon, the goose, the domestic fowl, the peacock, the parrot, the wagtail, the quail, and the "brain-fever bird" are reverenced, or feared. The parrot is a lucky bird to have in a house. Indian mothers will divide almonds between parrots and their small children, in order that the latter may acquire the parrot's fluency of speech. A quail is a lucky pet, because he attracts misfortune to himself. If a pigeon builds in the roof of a house, ill-luck will follow and the place will become deserted.

Vultures and kites are to be reckoned with. From a kite's nest the burglar obtains the magic stick with which he opens locks and doors. He secures the stick in the following manner: While the young birds are still in the nest, he fastens an iron chain to their feet. The mother-bird will then go and bring a magic stick with which to break the chain and release her young. After the escape of the fledglings the nest is taken to a stream, and the sticks of which it was built are thrown, one by one, into the water. The stick which moves off rapidly like a snake in the water, is the magic wand which the thief sought.

The crow and the owl are unlucky birds. However, food is given to crows in the belief that it will thus reach the pitris, or ancestors. A crow's caw in the morning signifies that a visitor may be expected.

The owl is a foreboder of evil. Still, it is dangerous to try to drive it away by throwing clods at it, for it may pick up a clod and rub it down to powder. In that case the thrower will fall into a decline and finally die, precisely when the clod has been reduced to dust. Both the owl and its flesh are used in magic.

Ants are sacred. They are worshipped with offerings of sugar, especially in May and June, and Chamars believe that they are able to answer prayers and grant children and other blessings.

Totemism is connected with the belief in spirits; and the life, or perhaps the soul, of some ancestor of the group which bears the totem's name was in some way associated with the totem. The names of totems found amongst the Chamars include those of trees, of seeds and grains, of birds, of animals, of individuals and of tribes.[1]

[1] Names of *gots*: (1) Named after trees: 'Ḍhākmaṭ (ḍhāk), Pīpaliyā (pīpal), Ambā (mango), Nīmgotiyā (nīm), Nīmoliyā (nīm), Kujariyā (date palm), Halduā (haldi), Simoliyā (cotton-wood). Other gots are named after the gular and the jhand trees.

(2) Named after birds: Parindiyā (generic), Chiriyāla (generic), Hams (goose), and Baṭeriyā (quail).

Some characteristic tabus are found in connection with the totems. For example, those whose gots are named after the gular, pipal, jhand, and nim trees will not cook their food with the wood of the particular tree that belongs to them. Those whose got is the *bheṛ*, will not eat the flesh of the sheep, nor drink its milk, nor use wool blankets.

Fetishes are common. Besides the stones of the village platform may be named the stone-mill, the pestle and mortar, the sil and batta, the plow, the winnowing-fan, the *khūrpī* (the hand hoe for cutting grass), the *rāṁpī* (shoemaker's or currier's knife), and the shoemaker's last. These all have their special uses, as illustrated in the customs described in the preceding chapters. Disease is treated by the drinking of water in which a fetish stone has been washed. The plow is garlanded on special occasions. The sieve is often the first cradle of the baby. The halter is a fetish of Jaiswar grooms. To insult this by tying a dog with it, results in a fine of five rupees.[1] The trident is often used as a fetish. The rings and chains used by the bhagat in spirit-control and the chains found in low-caste temples may be so considered. The *sanichar kā deotā*, the wooden beam of the plow, is another fetish. He comes upon a person on certain days, particularly Saturday, and causes him to cast the

(3) Named after seeds, grain and fruits: Maṭrī (pea), Siṁghaṛiyā (water-nut), Gutaliya (stone of mango fruit), Dhansaurā (rice in the husk), Masūriyā (a pulse).

(4) Named after animals: Bheddā (sheep), Suariyā (pig), Gidhariyā (jackal), *Bhaiṁsiya* (buffalo), Bheṛwāliyā (sheep), Achchhiyā bachhiyā (calf), Bardhiyā (buffalo), and Chheriṁyā (goat). Other *gots* are named after saints, gods, places, diseases, dust, etc., *e.g.*, Dhūliyā (dust), Koṛhiryā (leper), Kanhaiyā (Kṛishṇā), Kālīyā (Kālī), Dūdhiyā (milk), and Mukhtāriyā (strength). These names have been gathered from a wide field amongst the Chamars. Rose, in his articles on the Chamar, gives a number of Rajput clan names as names of *gots;* and Russell in his article gives a number of interesting names: Khumṭī (a peg), Chaṁdanīha (sandal wood), Tarwāria (sword), Borbans (plum), Mirī (chillies), Chauriā (a whisk), Baraiyā (wasp), Khalariā (a hide or skin), Kosnī (kosa or tasar silk), and Purain (the lotus plant).

[1] Crooke, *Tribes and Castes of the North-Western Provinces and Oudh*, Vol. II. p. 173.

evil eye. Such a person may be delivered from this state by being weighed with grain, iron and oil on a Saturday. At the special seasons of the Durga Puja and the Dewali the Chamar worships his tools and implements. Besides this somewhat individualistic, general attitude in the use of fetishes, where the man or family or the local group makes use of his particular implements or possessions, there is an emphasized personal use of the fetish for selfish purposes. The fetish is chosen because it is believed to be the habitation of some particular spirit or power. Unlike an idol, the fetish is not made to resemble the spirit; and unlike a god, the inhabiting spirit cannot occupy more than one object at a time. The fetish possesses personality and will, and may have human characteristics. The owner believes that the fetish may act by the will or force of its own proper spirit, or by the force of a foreign spirit entering or acting from without. So the fetish is worshipped, prayed to, sacrificed to, talked with, petted, and ill-treated. Offerings are made to it. The owner asks it or compels it to do his bidding. Professor Jevons[1] remarks that a fetish is private property, and that fetishism is anti-social and therefore anti-religious.

Nowhere is the Chamar's belief in spirits more clearly illustrated than in his superstitions about demons. These evil spirits are an object of propitiation. Their chief characteristic seems to be their incalculable nature which requires the "watch out" attitude on the part of the masses. It is especially the malignant dead whom the Chamar, by all means within his reach, propitiates. From the malevolent dead nothing is to be hoped for, but everything is to be dreaded. These evil spirits are more feared by women and by children than by men. Offerings of goats, pigs, cocks, eggs, grain, liquor, milk, water, and many other things, are made by way of propitiation. Besides, these ghosts require all sorts of prepared human food. The ranks of these spirits are recruited from the ghosts of the dead.

[1] *Introduction to the Study of Comparative Religion*, pp. 120, 121, 124.

Some say that any spirit may wander about for twelve months, and that one is never sure about them: they may be troublesome. If ghosts are still unsettled at the end of a year they become bhuts, if male, and churels, if female.

There are many kinds of demons and their names vary in different parts of the country. Names which are well known in some areas are almost unknown in others. But the general characteristic of these beings and the phenomena attributed to them are in all parts of the country the same.

The *Vetāl*, or *Baitāl*, the chief of demons, is described variously as wheat-coloured, white, or green. He rides on a green horse. He is sometimes counted as a godling.

The *Bhūt* is, in particular, the spirit of a person who has died a violent death, by accident, by suicide, or by capital punishment, or the spirit of one whose funeral ceremonies have not been performed. The bhut of Awadh is a tall, white, shining ghost who impedes men's progress along the roads at night. The term "bhut" is used also in a more general and comprehensive way to denote malevolent spirits.

The *Churel, or Churail*, is greatly feared. She is the ghost of a woman who has died while unclean, or while pregnant, or in child-birth; or, as some say, such an one who has died during the Dewali festival. She is described as having pendent breasts, large, projecting teeth, thick lips, unkempt hair, and a black tongue, and as of dreadful appearance. Her feet, like those of most evil spirits, are turned around. Some say that she is black behind and white in front. She is especially malignant towards her own family. To lay the ghost of a woman who has died as described above, and, consequently, to prevent her interference with the affairs of men, the body is sometimes buried face downwards, and some fill the grave with thorns and heavy stones to keep down the ghost. Again, small round-headed nails are driven through the nails of the forefingers and the two thumbs, and the great-toes are welded together with iron rings, to prevent the ghost from becoming active. The ground on which the woman died is carefully scraped and the earth removed, and the spot sown with mustard-seed (sarsoṁ), and mustard-seed is scattered along the road to the grave, or to the burial-grounds, to prevent her return home. Mustard blooms in the abode of the dead, and the Churel, who will stoop to pick up the seeds, is delayed until dawn, and then must flee. Sometimes a skein of thread is thrown into the funeral pyre, with the thought that the ghost will be taken up with the unwinding of the thread and so forget to return to trouble her relatives. Some burn the body to prevent the escape of the spirit.

Again, she appears as a beautiful young woman, seducing youths at night. She keeps them until they are prematurely old. At other times she comes in the form of a beautiful girl in white, and leads young men away to sacrifice. She appears in other forms too. An old Chamar wizard tells of two high-caste brothers, the stronger of whom slept in the fields at night to guard the crops, while the other remained at home. But the strong man suddenly grew weak and lean. Finally, his brother asked of him the cause of this great physical decline. In reply the other said: "A Churel comes to me every night and obliges me to cohabit with her." Thereupon, the younger brother decided to guard the crops. So, taking a pair of shears he went to the field. That night the Churel came and slept with him. In the night he cut her scalp-lock (*chuṭiyâ*) very stealthily and concealed it. In the morning the Churel awoke a mere naked woman, and she was unable to escape. So he gave her a loin-cloth (dhoti), took her home and kept her as his wife. They reared a family, and grandchildren were born to them. Then, one day, she asked her husband for her chuṭiya. She said that she wished to dance because she had grandchildren. At first he refused her request; but his friends agreed that, now that she had children and grandchildren to think of, she would not run away. So, at last, he granted her petition. As soon as she had obtained her chuṭiya she disappeared. The same Chamar related how a Churel came to him one night, about ten o'clock, as he was watching in a field. He thought it was his wife, and asked, "Who are you?" The Churel replied, "I am Bāba Dīn's mother." By this he knew that it was someone else, so he said, "Come along," made a place for her in his bed, and took out his knife so as to be able to cut her chutiya; but she discovered his design and fled. The Churel is often enrolled among the village godlings and given a place in the common shrine. All who see her are liable to be attacked by some wasting disease. And those who come out at night in response to her call are sure to die. If the Churel be seen in the home, a heated brick or a hot iron is thrown into the place that she frequents. She is thus driven away.

Another ghost, called the *Gayāl*, or *Ūt*, is the spirit of a man who has died sonless or unmarried, and who, consequently, has no one competent to perform his funeral rites. His malice is directed towards the sons of other folks, especially towards those of his relatives or his caste-fellows. The duty of performing his funeral rites devolves upon those next of kin or upon his neighbours, and they, in self-protection, see that these rites are faithfully carried out. In the Punjab, small platforms, in which are small hemispherical depressions, are constructed for the Gayal. In these milk and Ganges water are offered, and on these platforms lamps are lighted for him. A careful mother dedicates a coin to Gayal, and hangs it about her son's neck to protect him through childhood and youth, and until he has begotten a son.

The *Pret*, or *Paret* (fem. *Pretnī*), is the ghost of a deformed or of a defective person. To this class belong the spirits of those who were crippled, or who lacked an organ or a limb. The ghost of a child dying prematurely, or of a still-born child, may be called a Paret. The following narrative shows how vague may be the conception of just what a Paret is: "On one occasion I was with a Brahman in a field where sheaves of grain were piled beside the threshing-floor. Some time after dark I noticed that some of the sheaves were being thrown about; so I suggested to my companion that he drive away the animal that was causing the trouble. But he refused to do so. Thereupon I took a club and went to drive it away. I saw a bullock, which, as I stood and looked, changed itself into a horse. The horse became a camel, and the camel became an elephant. I had presence of mind enough to call upon my *Bīr* (a powerful demon) for help. The Bir came, and when I knew that he had come—by a peculiar twitching in the flesh of my right upper-arm—I felt safe. The elephant increased in height to about thirty feet and then disappeared. I was taken ill with vomiting and diarrhoea, but relief came after my father had made an offering to the Bir." The same narrator told the following story: "During one season, while I was watching the fields at night, I slept under a tamarind tree. Every night an evil spirit came and lay down on my chest. After this had happened a number of times, I spoke to a Brahman, who was sleeping in an adjoining field, and to his uncle. They both laughed at me. So, when the evil spirit came upon me again, I spoke to him, saying, 'If you are able to do so, go to that Brahman.' The ghost went that very night. Later, I heard the Brahman call out, 'Ah! Ah!' I shouted to him five times, calling him by name (Rāmapat). He was unable to answer immediately, but a little later he called out to me, saying, 'Did you call me five times? I heard you, but I could not speak because someone was sitting on my chest and holding me by the throat.' The Brahman fell ill and died." Such stories as these show how completely the Chamar lives in the fear of evil spirits. The Pret is not always malicious.

The *Pisach* is a demon resulting from a man's vices, and is in reality the spiritual embodiment of some vice, as the lying spirit, the thief spirit, and the like, or the spirit of insanity.

Another much dreaded demon is the *Masān*. He is especially ill-disposed towards children, whom he often changes to yellow, red, or green colour. He also causes them to waste away and die, by casting his shadow upon them. He is known only by his works, and, because of his invisibility, is most dreaded. If water from the cooking of the food fall on the fire so as to put it out, the household is in terror lest the children be beset by Masan. If a woman allow her chadar to drag behind her the Masan will follow her home. He will not disturb her, but the children will pine away. And if children are born they will die. After putting out a lamp with the fingers, it is unsafe to rub them on the clothes for fear of the Masan. A child may be delivered from the power of the Masan by being weighed in salt. The Masan is said to be also the ghost of a child, and that of a low-caste man (a *telī*, or oil-presser).

The female demon, *Masānī*, is the spirit of the burial-grounds. She comes out at night from the ashes of the funeral pyre and attacks people as they pass by. The Masani is black and hideous in appearance. She is often rated as a sister of Sitala.

The *Rākshasas* are ogres, or giants, found in trees, in birds, and in cisterns. Some are deformed. They sometimes animate dead bodies. They devour human beings, and eat raw flesh and carrion. They cause vomiting and indigestion. They carry under their finger-nails a deadly poison. They often assume the form of an old woman with long hair. When they take human form their heels are in front. Local tradition often considers them as the architects of ancient buildings now in ruins. Like other demons they are active at night, when they mislead travellers. They are easily fooled, and can be made to disclose their secrets. They travel through the air, and depart with the dawn. Among the especial classes of rakshasas are the *Deo*, a gigantic, powerful, stupid, long-lipped cannibal; the *Bīr*, a malignant village-demon of great power, who, amongst other things, brings disease upon cattle; and the *Dāno*, who often lives in a bargat tree. The wizard in whose house the Bir lives may ask what he wishes and the demon will carry out his request. But the Bir may live anywhere and still be in the wizard's power. The wise man may summon him at will. The Dano often pounces upon men, especially young men, at night.

Another dreaded demon is the *Ḍūṇḍ* (or truncated), the ghost of an unburied Mussulman martyr. He rides a horse, but has neither head, hands, nor feet. He has his head tied on to the pommel of his saddle. He comes periodically, and calls out to people at night; and he who comes out-of-doors in response to his calls is sure to die or to go insane. Frequently rumours are afloat that the Dund is about, and then all people keep carefully indoors at night.

There are a number of demons with generic names, as the *Brahma-Purusha*, the ill-tempered ghost of a Brahman; the *Baramdeo*, a similar ghost; the *Manushyadeo*; the spirit of a widow's deceased husband; and the spirit of a second wife's predecessor. These spirits must be given plenty of attention. The *Bhagaut* is the ghost of a man killed by a tiger.

Among the fiends are the *Chordeva* (sometimes called the *Manushdevā*), *Jilaiyā*, *Raṛuīchiryā* and *Marelī*. Chordeva is a birth-fiend, who comes in the form of a cat and worries the mother or tears her womb; so cats are not allowed in the birth-chamber. Jilaiya takes the form of a night-bird and sucks the blood of persons whose names it hears. So children are not called by name at night. If this fiend fly over the head of a pregnant woman her child will be born a weakling. The Mareli is a bird-fiend, who comes and sits on a tree near a house where a man lies sick, and calls out. If anyone should throw a clod at her, she picks it up and drops it into a tank, or pond. As the clod dissolves, the sick man wastes away and dies. If this bird is killed on a Sunday and its body burned, after certain incantations have been pronounced over it, the ashes become a valuable love-charm. Any woman over whom a man throws these ashes will follow him.

Pherū and *Rahmā* are now demons of the whirlwind. To avoid the effects of an approaching whirlwind a person should repeat the charm: "*Bhāī Pherū terī kār*" (I am within thy charmed circle, O holy Pherū). If this formula be repeated three times, the evil spirits who come with the whirlwind will do no harm. There are demons of the storm, of the lightning, of the thunder, and of other natural phenomena.

The *Dodo* and the *Hawa* are invoked to scare children. Other bugaboos abound, one of whom is an old man with a boy who carries off naughty children.

The number of demons with functions and characteristics like those described above is legion.

The *Paris* are fairies, most beautiful spirits, who carry away beautiful persons. They take away the blossoms of the gular tree at night. These are for the most part creatures who are harmless, and who fall in love with human beings. They are visible to the pure eyes of childhood. On the other hand, the Paris attack men on moonlight nights, catching them by the throat, half choking them and knocking them down. They protect children.

The Chamar accepts also most of the Mohammadan varieties of spirits, such as *Jinns*, *Ifrīt* and *Marīd*.

All demons require food, preferably the blood of animals, and they must be propitiated; yet they have no regular worship and no imposing temples.

Since demons multiply in proportion as the death ceremonies are neglected, everything is done to facilitate the passage of the spirit to the abode of the dead, to "lay" the ghost, and to "bar" its return. So the dying man is placed on the ground, and the mourners at the funeral wail to keep off evil spirits of obstruction; the body is carried feet foremost, and other devices are employed during the funeral ceremonies as provision against any possibility of the spirit's return; and here again we see the significance of the funeral rites. The burial-party bathe after the cremation or burial ceremonies are completed; on the homeward journey they do not look back; on the way back they step over running water and throw bricks or stones over their shoulders; and when they reach the house, they shake out the folds of their garments and touch stone, cow-dung, iron, fire and water; then they also touch the left ear with the little finger of the left hand, chew nim-leaves, sit in silence to allow the spirit if it has come so far to depart, and then disperse in silence. The chief mourner carries a lota about with him until the funeral ceremonies are completed.

There is a close connection between disease and demons, and many kinds of sickness, and often death itself, are attributed to demoniacal influences. This close association in the minds of the Chamar is emphasized by the fact that most of the means used to scare demons, and as protection against evil eye, are used in the prevention and cure of disease. The demons that cause disease are legion. Their worship is of the crudest form. Only in time of calamity, or when epidemics are rife, is much attention paid to them. Long periods of health and prosperity result in the neglect of these godlings and in their shrines falling into decay. These demon godlings, whose displeasure brings disease, are more or less local. Of course, any demon may be responsible for the disease. In most cases of illness the demon responsible for the trouble must be identified by the *Sayānā*, or devil-priest. Some diseases have caste names, as, for example, one form of smallpox, *Chamāriyā*.

When an epidemic is raging, all the powerful disease demons and malignant godlings are propitiated.

The line of distinction between godlings and demons of disease is hard to draw. Many of them are known as forms of *Kālī*. The names of aspects of Kali which are current in some parts of the country are scarcely known in other areas. She has control over many forms of disease, among which plague is now prominent. If propitiated she will prevent disease, but if angry she will bring it upon those who have offended her. Her power is felt in diseases other than smallpox, although Sitala is sometimes considered as one of her forms. It is to this dreadful Kali, in some of her aspects, that offerings are made when epidemics are raging.

Marī, or *Marī Māī*, the cholera goddess, has special shrines, and the nim tree is worshipped as her abode. Offerings of pumpkins, cocks, male buffaloes, rams, he-goats and puris are made to her. The offered animal is decapitated at one blow before her altar. *Umariyā Mātā* is worshipped in cases of cholera. The word "cholera" (*Haiza*) is sometimes personified. The dread disease is often attributed to some unnamed but powerful demon. In some places a new plague-goddess has appeared, *Kaṁṭhī Mātā*.

Sītalā Mātā (Mātā, Mother; Jagrānī, Queen of the World; Mātā Māī, Great Mother; Jagadamba, Mother of the Earth; Kaḷejewālī, She of the Liver; Ṭhaṁḍī, She that Loves the Cool; and Phapholewālī, She of the Vesicle), the goddess of smallpox, lives in the nim tree. Her feast day, which is known as Sītalā kī Saptamī, is the seventh of the dark half of each month. No fire is lighted then. She is also worshipped on each Monday in the month of June. Her worshippers are women and children, never men. As a household goddess she is called Ṭhaṁḍī, and she has her place behind the waterpots, where she is worshipped by the house-mother with cold food and cold water only. She has special shrines and small temples, sometimes in charge of a devil-priest, or of a low-caste man, a Chamar or a sweeper. There is one to her in the Muzaffarnagar district, where she is worshipped as Ujālī Mātā, or the Bright Mother. The offerings made here are cakes, sweetmeats, and gur. When children are suffering from the disease, water is poured over her shrine. This is magic. Another shrine, at Rāewālā in Dehrā Dūn, is visited by large crowds. Here vows are māde to obtain children, and children born by her gracious favour are brought to the shrine. Offerings are made in lives. There is a *thān* (shrine) of hers at Sikandarpur, in Bijnor, where a mela is regularly held. There is a temple in Gurgaon, open to all castes, where special religious fairs are held and where every Monday people come to worship. There is another temple at Jalaun. These notes illustrate the local character of her places of worship, which are found all over the country, nearly every village or group of villages having such a place.

Sitala carries a broom and a basket. She sweeps men about when she comes, gathers them in her winnowing-basket and scatters them to the winds.

Her vehicle, the donkey, on which she rides in a state of nudity, is a type of slow motion, which means that she takes a long time to go away. Some say that she rides on a tiger.

She is one of seven sisters, who are supposed to cause pustular diseases. One is Masani, who plagues people with boils. She has ears as large as winnowing-fans, projecting teeth, a hideous face, large eyes, and wide-open mouth. She rides on an ass, carrying a broom in one hand, a pitcher in the other, and a winnowing-fan on her head. The offerings made to her are afterwards given to scavengers and jogis. *Agwānī* is the fever-goddess, who heats the body. Sitala's elder sister, *Chamāriyā*, is the disease in its worst form. This is an interesting name. Other castes make offerings to her in the form of a pig sacrificed by a Chamar. Her younger sister, *Phūlmatī*, represents a mild form of smallpox. The other sisters are *Basantī* and *Lamkariyā*.

A peculiarity in the case of Sitala is that the disease is the goddess, and the eruptions are signs of her presence. Some say that "during an attack of smallpox no offerings are made, and, if the epidemic has once seized upon the village, all worship of her is discontinued until the disease has disappeared. But, so long as she keeps her hands off, nothing is too good for the goddess."

A considerable body of magic has grown up about the treatment of smallpox. When the dread disease is about, the *chapātī* is not cooked in the usual way. When the loaf is half kneaded, it and the cook's hands are dipped in flour before it is flattened. If this is done, blisters do not form when the bread is placed on the tawa to be cooked. Besides, this method obviates the sputtering sound, which is offensive to Sitala, and so averts her anger. Another valuable practice is to use simple and unusual methods of preparing food when smallpox is in the village or mahalla. So the food to be cooked is put into the pot all at once. During an outbreak of the disease, women worship Sitala's shrines and pour water over them to keep her cool. Water is poured also at the foot of the nim-tree and at cross-roads.

In Tirhut a feast to Jūr Sītal, or smallpox fever, is held. The people bathe in water drawn the previous night, and eat cooked food after worshipping her. From morning till noon they cover their bodies with mud, and throw it over all whom they meet. In the afternoon they go out with clubs and hunt jackals, hares, and any other animals that they happen to meet in the village. After they return they boast of their valour.

Women visitors are not allowed to come into the sick-room, and while the disease is raging people will not go on a journey, not even on a pilgrimage. Offerings of flowers, milk and Ganges water are made. For relief during the course of the disease seven suits of clothes, bound in a thread, and betel-nut, are waved over the patient and then cast into a well. The black dog is respected and fed as a propitiation when smallpox is about, and sometimes a donkey, the vehicle of Sitala, is fed with fried gram. Gram is also waved over the head of the sick child, presented at a shrine, and given to the donkey's master. Fowls, pigs and goats are offered. A white cock is sometimes waved over the patient and then let loose. As thunder disturbs Sitala, the stone mill, or copper plates, or cooking utensils are rattled near the child's ear during a storm. If the smallpox disappear prematurely, a relative goes at night to a tank, naked, and brings, in a new vessel, water from beneath a dhobi's washboard. Some of the water is then poured over one of Sitala's shrines, and the rest is brought home, passed into the house through the roof from behind, and then sprinkled on the patient. Sometimes, after six or seven days, a sick child is covered with silver leaf and given raisins to eat; and, as the disease abates and the pox dries up, water is sprinkled over the body of the child. Musicians are called in, and the child is dressed in saffron-coloured clothes and carried to one of Sitala's shrines. There a pipal tree is besmeared with red-lead and sprinkled with curds, and red rags are tied to its branches. Thorns are cast in the pathway leading to the infested place to bar her return. Other elements in the treatment of the disease will be described under folk remedies and magic.

When an adult has recovered from smallpox a pig is let loose in the name of Sitala, lest the patient have a relapse. Upon recovery, offerings are made, consisting of cocoanut, betel-nut, haldi, dub-grass and a black goat. Smallpox must have been a most terrible scourge in the days before vaccination was introduced.

Strange and impracticable as it may seem, there are unfailing tests by which demons may be recognized. They cast no shadows; they can stand almost anything but the smell of burning turmeric, and they always speak with a nasal twang. Some are deformed, while others are of special colours. There are many places where bhuts are likely to be found, and where it is unsafe to venture unless well protected. Burial-places, cremation-grounds and all deserts are infested with demons; birds such as the owl are possessed by evil spirits; empty houses and old ruins are haunted; some ancient ruins are attributed to demoniacal activity; the sites of old villages are respected as infested with ghosts; mines and caves are abiding-places of demons, and evil spirits are said to be guardians of hidden treasure. It is because spirits frequent cross-roads and highways that sometimes, to get rid of disease, a stake is driven into the ground at the crossing of roads, and seeds are scattered about it, and that smallpox scabs are placed at road-crossings and along the highways. The village boundary is a place where all sorts of demons congregate, so we have *Chāmuṇḍā*, or *Sewānrikī* (*Sewanriyā*), taking up her post to protect the village from foreign spirits. Foul places are the abodes of demons, so the Chamar cleans his house but leaves his yard filthy. Demons are found on the roofs of houses. But bhuts can never sit on the ground (Earth is a devil-scarer); hence, at low-caste shrines, pegs or bricks are set up, or a bamboo is hung over the shrine as a resting-place for demons. A person who is going on a pilgrimage to the Ganges, bearing the bones of a dead man, sleeps on the ground but hangs up the bones, as they must not touch the earth. Near shrines it is best to sleep on the ground. The dying man is placed on the ground, and the bride and groom sleep on the ground. Sweet-smelling flowers are infested by bhuts, and for this reason children are sometimes not allowed to smell them; nor do the people use perfumes on children. Demons are very fond of milk, and so that must be protected by a piece of charcoal. They are never found in the temples of the gods, although they are always near by.

Spirits attack and enter the body through the head, hair, mouth, eye, ear, hand or foot. Some say that the Pret enters through the feet, the Deo through the head or hands, but the Bhut through the eye or ear. So the feet are washed at weddings, and the bride is lifted over the threshold; the head is shaved at puberty and at times of mourning; the eyelids are blackened; bracelets and anklets are worn, and many other devices are made use of to outwit demons. Opportune times for demons arise when one yawns, so one should at such a time clap his hands or snap his fingers and call out "Nārāyan!" At meals care should be taken in preparing food and in eating, and during festivals; hence the elaborate preparations in the domestic ceremonies. The effects of spirit entry, or possession, are disease, barrenness, loss of favour or of affection, failure in business, and general misfortune.

The times when persons are most subject to demoniacal possession are at birth, at marriage, and at death, the great crises in life; and, consequently, it will be perfectly clear that many of the ceremonies connected with these events have for their object the scaring of ghosts. Besides, women and children, as in the old classic days, are always subject to demoniacal influences. It becomes necessary, therefore, to work out a system by which demons may be kept off, and means by which people may be protected from evil influences. The following devices apply to spirits in general, but they all have in mind evil spirits in particular. The bridegroom wears a crown and clothes of bright colours; both the bride and groom are protected by wave ceremonies and many other devices. The foods, the grains, and the colours of various ceremonies are chosen for protective purposes. The tuft of hair furnishes a resting-place for a spirit, so sometimes the added precaution is taken of tying into it a piece of blue-black thread or rag. Lampblack is rubbed on the eyelids. Rings and bracelets are used. Charms and amulets are worn. Women put *meṁhdī* on their hands and feet. At night, when travelling, a Chamar meeting another person will not speak, lest, his voice being heard, he draw upon himself some evil influence and his business come to naught.

An enumeration of the various devices used as protection against demons throws a flood of light upon the commonplace practices and customs of the Chamar.

Iron is potent in keeping off demons, especially when it is fashioned into a tool. Horseshoes are found at shrines, and they are nailed to the threshold to keep out evil spirits. Iron is found in the bed in the lying-in room; the mother wears an iron ring during the days of impurity after childbirth; nails are driven into bedposts, and are used in "laying" the Churel; and he who lights the funeral-fire carries iron. The cooking-vessel is turned upside down at night, at the right of the head of the bed in the lying-in room, to protect the child from demons. If nails are driven into the four bedposts no evil spirit will attack him who sleeps upon the bed.

Other metals are in constant use: copper in rings, amulets and coins, and brass in the lota, which every mourner carries with him until the death ceremonies are completed, and which is used when a person goes to perform the offices of nature. Bell-metal and other mineral products are worn as bracelets, anklets and rings, and around the neck. Tinsel is found in the crown and on the clothes of the bridegroom, and in the kaṁgna.

Marine products, such as coral and shells, are used as ornaments, but with a practical purpose as well. *Kauṛīāṁ* are often seen bound on the arm, or around the neck, or in the hair; and they are used to protect animals as well.

Precious stones, while believed in, are beyond the reach of the poorer classes. Imitations are in use. The cutting and the shaping of these is significant. Much use is made of beads, especially blue beads. Rosaries and the *hār* have protective powers.

Salt is a potent protection. Salt and red mustard are scattered around tbe patient's head to cast out fever. Salt is given after sweats, to protect children from the Churel.

Incense and smoke resulting from burning certain things are also potent in keeping off devils. Incense is burned at the chhatthi ceremony. Sometimes, when a child is ill, incense is obtained by burning bran, powdered chillis, mustard and the child's lashes. While these ingredients are burning the fire is waved around the child's head. In most instances the vileness of the smell of the incense is the important thing, and the worse it smells the better the result. Among other things that are burnt for similar purposes are leather, human filth, tiger's flesh, human hair, droppings of pigs and dogs, sulphur and *lobān*. Cow-dung and many kinds of filth are used as devil-scarers, and the five products of the cow are most potent.

Blood, especially menstrual blood, is a potent charm against demoniacal influences, except those of the Churel. Traces of the use of blood are seen in the red marks made upon the drum in certain ceremonies and in the marriage ritual, where vermilion is used for the tika, and in the handprint on the houses.

Water, fire, earth and ashes all find an important place in social and religious rites where protection from evil spirits is necessary. This explains, to some extent at least, the ceremonial bathing in the domestic customs, the use of lights and of the fire-sacrifice, the magic-earth ceremony, and numerous other devices of a similar nature. Water from a tanner's well is very effective.

Grains, particularly barley, rice and urd, as well as mustard, are used. Parched grain is valuable. Special spells are pronounced over rice. Among spices and vegetables turmeric is to be especially noted; likewise the betel-nut, the cocoanut, the plantain and garlic.

Many colours are regularly used, chiefly yellow, red, white, blue and black. The wedding-garments are of red and yellow; the woman during her period of pregnancy uses blue-black clothes or threads; and coloured threads enter into the domestic ceremonies. Charcoal is put into milk, and lucky signs are made with it (and in red and in cow-dung) on doors and pots.

Oil and ghi are used on images and in the ubtan. They are often mixed with red-lead.

Feathers are used, as in the worship connected with Zahra Pir. A bit of peacock-feather, struck on the wall above the waterpots with cow-dung, protects the drinking-water from the evil eye.

Branches and leaves of the dhak, the nim, the bamboo, the castor-oil plant and the tulsi shrub are potent scarers. Grasses are also used.

Leather, especially shoe-leather, is a devil-scarer; so the father puts his shoe upside down at night, near the foot of the bed, as a protection for his child. Shoe-heels are also used. When the spindle of the spinning-wheel begins to "talk" and does not run well, they beat it with a shoe to drive out the demon who is making the trouble. When a person begins to scratch his nose it is believed that he will be attacked by some disease due to a demon. It is wise, therefore, to take a shoe from some person who comes into the house, and rub the nose seven times with it as a preventative measure. To cure epilepsy the sufferer is made to smell an old shoe.

Some special devices used to drive off demons include wave ceremonies, noises, and figures. Waving scares demons. This partly explains the wave ceremonies mentioned above, and accounts for flags at temples. Nim branches are waved to exorcise spirits. If, at the time of the wedding, the activity of the power of fascination is suspected, salt is waved around the head of the bride and groom, and burned near the house-door as a charm. Certain forms of dancing may have this significance. Noises scare demons. This accounts for the use of loud and noisy music at festivals and ceremonies and in many practices connected with demonology. The beating of the tawa and the ringing of bells belong here. The clapping of one's hands, as noted above, has this function. In visiting an old tomb one may catch a harmless spirit unawares, and, as he resents being disturbed, one should clap one's hands. And yet silence is maintained at special times, as during the measuring of the grain. Certain symbols and figures are of great value. They are found about the house and on the cooking-vessels, upon the pots and about temples and shrines. The circle in various devices is found everywhere. It appears in rings, in kaṅgnas, and in other ornaments. The magic circle is drawn in connection with the Nag Panchami festival. The circle is found also in knots and in arches. Both the double and the single triangles and the square are common. The swastika is found upon doors and door-posts and in many other places. This is to be seen where family gods are placed, and often in a shrine of Bhumaiya. Here it consists of two straws on a daub of cow-dung plaster. It is a symbol of blessing. The magic hand appears in red, yellow and black, and in cow-dung. When found upright on a shrine, it denotes a prayer; when reversed, it indicates that the prayer has been answered. The protecting hand may be seen on bullocks and on houses. Other references may be found in the domestic customs. Crossed lines are used. Various figures are used in connection with the worship of Shasti, and many other symbols are drawn for protective purposes.

Connected with the subject of protection from demons is the use of caste-marks and tattooing. Some Chamar women wear on the feet, as a distinguishing caste-mark, a specially shaped anklet, called *dhundhnī*. It used to be the custom for Chamars who were at work with other caste-men, and who did not wish to conceal their caste, to tie around their pipe (chilam), or tongues, a small leather thong. Chamar women, much more than other women, have themselves tattooed. This is done upon the breast, stomach, upper and lower arms, hands and feet, and upon certain parts of the face. In some cases this is a ceremony of initiation. These are the only ornaments that a Chamari can take with her beyond the grave. When a Chamari dies Parmeshwar asks her to display the marks and signs which she ought to possess to show that she has lived on the earth. If she cannot show these tattoo marks, she will not see her father and mother in the next world, but will reappear as a bhūtnī, a pretnī, or a rākshasī. So tattooing takes place about the time of the marriage ceremony.

The origin of the use of many of these devices for scaring ghosts is undoubtedly in utility. The medicinal value of certain herbs and of the leaves of certain trees explains some of these practices. Other are more obscure.

Many are the hit-or-miss results of insufficient observation and have been handed down from time immemorial.

Some of the devices described as devil-scarers appear in the practice of magic also, and in connection with the belief in the evil eye. And these two latter phases of primitive belief and practice illustrate further the methods used to control, or to defeat, the plans of demons.

The materials presented so far in this chapter, and a good deal of material in the preceding chapters, show how thoroughly the attention of the Chamar is occupied with malevolent spirits. He has, however, a great number of benevolent spirits as well, whom he enlists for protection and for aid against the forces of evil. The sainted dead and other spiritual beings, some of whom have attained unto some degree of divinity, and the godlings occupy an important place in his thinking. It is difficult to draw the line in some cases and to determine whether the ghost is a benevolent or a malevolent spirit. If some person die under unusual or untoward circumstances, or if some extraordinary event transpire, a shrine is built to appease the spirit concerned. So, special shrines and platforms are constantly appearing. Many sadhus and other holy men are revered, and their worship is carried on over a more or less wide area for a time after their death. In these ways countless local shrines arise, to which pilgrimages are made in the expectation of material prosperity, relief from disease, or the boon of offspring. Examples of such places are two graves of Nanak Panthi gurus at Bhogpur (in Bijnor) and that of a sadhu at Jhalu (in Bijnor). The graves of the former are visited in the rainy season. At Bijnor two brothers, Nur-ud-din and Shahab-ud-din, are rated as saints. To one batasas, and to the other cups of spirits are offered. In times of sickness these brothers are worshipped, and in their names vows are made. Their graves are visited five times a year. At each of the first four visits the worshipper walks around the graves once, and on the fifth visit five times.

In a Chamar village, in the eastern part of the Provinces, there is a large beehive-shaped shrine to Hem Rāj. The structure, which rests under a tree on the

THE POT MARKS THE POSITION OF A VILLAGE PLACE OF WORSHIP

PLACE OF WORSHIP, WITH OFFERINGS, NEAR THE VILLAGE

SHRINE OF NAT BABA

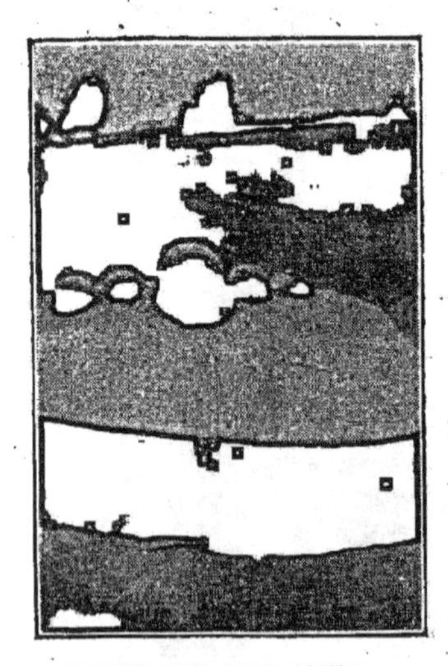

KALKI AND HER COURT

SHRINE OF HEM RAJ

outskirts of the village, is in a dilapidated condition. In front may be found a small earthen saucer with a wick in it, the remains of a lamp placed there as an offering. One day, while Hem Rāj was entertaining the people of the village, a woman cast a magic spell upon him. He fell down in a fit of apoplexy and died. He was buried on the spot, and a mound of earth was erected to his memory. Here worshjp is still carried on.

The legend of Nat Bāba also illustrates how places of pilgrimage arise. On one of the banks of the Ken river is an old and ruined fort, and back from the opposite bank is a rocky hill. In the old days, the Raja who owned the fort challenged a Nat (one of a tribe of wandering acrobats) to stretch a rope from the hill to the fort, while the river was in flood, and to walk on it from the hill to the fort. The Nat was to receive half of the kingdom should he be successful. The challenge was accepted and the performance begun. When the Nat reached the middle of the stream the Rani began to fear that the Raja would lose half his kingdom. So she tried to get someone to cut the rope and so plunge the Nat to death. No one, however, was willing to cut the rope, because the Nat was performing his feat by means of magic. At last it was reported that the Chamar's rampi was a non-conductor of magic, and a Chamar was persuaded to cut the rope. Before he died, the Nat cursed the queen and the kingdom. The ruined fort is the fulfilment of his curse. The Nat was buried below the fort, on the river-bank. His grave is now a place of pilgrimage, where people make offerings for children and wealth. The Nat has become a saint and is called Nat Bāba.

In some parts of the country a group known as the Five Saints (*Pāñch Pīr*) are revered. The names of the five are not always the same and many are found in the list. We have in these a grouping according to a lucky number, five. They are better known towards the Punjab. The names of the five are, for the most part, of local significance, and many of them are names of Mussulmans. For these five Chamars set up five pegs in their homes. Among the five are (1) *Shekh Sarwār*, who is buried at Hardwar. He is worshipped with offerings of very thick bread and coarse flour. (2) *Mirān Sāhib* is a headless horseman who has a dargāh at Amroha. He has a brother, (3) *Giar Samadān.* Both of these brothers have shrines at Bijnor. (4) *Gāzi Miān*, who died at Bahraich in 1034, in early life, is sometimes reckoned amongst the five. (5) *Gāzi Sālār* (*Bara Miyān, Bāle Miyān, Masūd Sālār Gāzi*) is now the patron saint of the inhabitants of the British cantonments of North India.[1]

Another saint, *Kalu Bīr, Kalu Bāba*,[2] was a brave strong man of the Gūjar caste. His grave is at Barhapura (in Bijnor), where he has many followers. Tradition has it that he is the son of king Solomon and a Kahar girl, who by magic compelled the king to marry her. The saint has a good many followers, especially in the Meerut Division, many of whom are Chamars. His fetish is a stick decorated with peacock feathers, and he is worshipped with petty offerings of food. It is said that, if gur and cakes are offered to him, he will lift wagons out of ruts and do other similar tasks that require great strength.

Būṛhā Bāba (*Bābu*) was a dwarf of the potter caste. Some say that he was only three feet tall, but that he was so large of girth that his belt would enclose twelve buffaloes. If he is not properly propitiated he will cause white leprosy and other terrible diseases. But he protects and serves his friends. When disease is epidemic and the cause is attributed to Burha Baba, a potter is summoned. Under his direction the sick man's friends take clay from the potter's wheel and apply it to the diseased parts. Then offerings are made, some of which are set apart for the saint, and placed before the potter. In connection with the preparation for the wedding Burha Baba is worshipped. A visit is made to the village potter, where his wheel is worshipped with offerings of grain and marked with haldi. Afterwards the vessels for use in the wedding are purchased

[1] *Legends of the Punjab*, Vol. I. No. V. and notes.
[2] See Chapter VIII. p. 219.

and taken home. The potter comes to the house and makes a kaṁgna out of a strip of new white cloth, by fastening in it a betel-nut, an iron ring and a bit of turmeric. This he ties on the wrist of the bride (or groom) for protection. In Burha Baba we have a good illustration of the mixture of elements belonging to both the fear of the malevolent dead and reverence for the benevolent dead.

Another interesting saint is *Bāba Farīd*[1] (Baba Shekh Farīd), a famous robber. One day, when he was about to rob and murder a faqir in the jungle, the saint asked him who of his family would go surety for him in this world and in the next. Farid asked his family and they all refused to do so, so he reformed. Baba Farid is also known as *Śakkar Ganj*, or *Ganj Śakkar*, from the reputed honour of having turned stone into sugar. He was a thrifty saint, who, for the last thirty years of his life, nourished himself by holding to his stomach wooden cakes and fruits whenever he was hungry. He had a magic bag from which he could get anything he wished. In the upper Doab, the ceremony of the first boiling of the sugar-cane is connected with him. Sugar-cane juice is passed around, and then from the first of the gur five cakes (bheli) are set aside for the Five Saints, of whom they reckon Baba Farid as one. They are left until the work of making gur is completed, after which they are distributed. Out of the first of the gur some is passed around also, and this is called *farīdī*. Should the saints be neglected, they would bring a curse upon the sugar, and there would be no profits.

Gorakh Nāth,[2] the famous saint and ascetic and worker of miracles, is recognized by the Chamars in some

[1] Crooke, *A Rural and Agricultural Glossary for the North-Western Provinces and Oudh*, p. 94; also, *An Introduction to the Popular Religion and Folklore of Northern India*, p. 135.

[2] On Gorakh Nāth see Temple, *Legends of the Punjab*, II., VI., XVIII., XXXIV., XXXVIII. and LII.; the article "Kanpatha" in Crooke's *Tribes and Castes of the North-Western Provinces and Oudh*, Vol. III. p. 73; *Encyclopædia of Religion and Ethics*, Vol. VI. pp. 328 ff.; *Indian Antiquary*, Vol. VII. p. 293; XXIV. p. 51.

areas as a saint. He possessed a magic wand, and wooden sandals that conferred wonderful powers of locomotion. Any person to whom he gave the sandals was able to fly. In some legends he is spoken of in connection with *Bhīmsen*, one of the heroes of the Mahābhārata. It is related that when Bhimsen lay benumbed with cold on the snow-covered mountains, Gorakh Nath restored him and made him king of 110,000 hills. To these two saints is attributed the substitution of buffaloes for human beings in sacrifice. Sadhus of this sect, who are called Gorakh Nāthīs, or Gorakh Panthīs, or Kānphatas, wear very large earrings (*mudra*), and have a miniature horn or whistle (*swagi*) hung from their necks. This whistle is used in worship and in ceremonies connected with bathing and eating. Some of them wear rosaries made of beads of stone secured on pilgrimage to Hinlāj, in Baluchistan. They practise some revolting forms of austerity (*yoga*). Householders of this panth, some of whom are of low-caste origin, are not admitted to holy orders. The great sanctity which Gorakh Nath possessed enabled him to do many wonderful things. His name is constantly mentioned in the legends of *Sakhi Sarwār* and of *Gūgā Pīr*. This circle of legends deserves some consideration because the names are widely known. But as one travels to the east he hears less and less about Sakhi Sarwar and Guga Pir. These legends illustrate what is in the minds of all classes of persons in connection with the veneration of saints and the pilgrimages to shrines. The shrine of Sakhi Sarwar Sultan,[1] at Dehra Ghazi Khan, is a celebrated place. Since he is especially benevolent in the granting of sons, many village women in the Punjab are his followers. In the Delhi territory he has shrines to which pilgrimages are made, and where vows are made in anticipation of the boons which he is able to grant. Attendants as well as pilgrims all sleep on the ground, and there are no beds in the adjoining village. He had a famous mare, Kakkī.

[1] See Temple, *Legends of the Punjab*, Nos. II., IV., VII., VIII., XXI., XXII., XXXVIII., XLV. and LIII.

Rāja Bāshak, a godling of the under-world, is the king of snakes. He is celebrated in the legend of Lona Chamari and in those of Guga Pir. Bashak is able to appear as a snake or as a man, and he uses snakes as the medium of his power; but he is under the control of Guga. Snakes are called Guga's servants. Guga is now, in the Punjab, the greatest of snake-kings. It is reported that he was found in his cradle sucking a cobra's head. The saint Guga Pir, or *Zahrā Pīr*, was born a Hindu; but he afterwards turned Mussulman, in order that he might enter the interior of the earth and bring the snake kingdom under his control. He is well known in the western parts of the Provinces and in the Punjab; and he has shrines far to the east, although he is less known there than in the north-west. The legend[1] of Zahra Pir, or Guga Pir, is one of the most famous in Northern India. He is worshipped to prevent snake-bite and in cases where persons have been bitten by poisonous snakes or by scorpions. It is good to listen to the story of Guga at night during Dewali, since the mention of his name deters snakes from entering houses. When vows made to Guga are not fulfilled, it is believed that a snake appears in the house within twenty-four hours and demands the gifts within a certain specified time. Some of his shrines are still famous. At one, in Multan, cures are wrought for blindness, barrenness and leprosy. There is a special festival, known as the *Chharī* (*chhariyā*) *melā*, held during the rainy season in honour of Guga Pir, which is very popular amongst the low-caste people in the north-west. This fair is named after the chhari, or flagstaff, which is carried in his name. Among the things necessary for the worship of Guga is the "flag," which consists of a bamboo twenty or thirty feet in length, surmounted by a circle of peacock feathers, and decorated with fans and flags and

[1] See *Indian Witness*, Feb. 21, 1911; Temple, *Legends of the Punjab*, Nos. VI. and LII.; *Indian Antiquary*, 1895, p. 49; 1897, p. 84. Crooke, *An Introduction to the Popular Religion and Folklore of Northern India*, p. 133; Oman, *Castes, Customs and Superstitions of India*, p. 67. *The Chharī Kā Melā*, by the Rev. A. Crosthwaite, in *The S.P.G. Mission Quarterly Paper* (Cawnpore), Jan. 1910.

cocoanuts done up in cloth. At the fairs men who are called "Zahra Pir's horses" carry these "flags." The poles are also kept at home by some persons and are used in special sacrifices. They are sometimes carried from house to house in August and then the owners receive alms. The object of the mela is to do special reverence to Guga, and to insure thereby immunity from snake-bite. Often at the foot of the flagstaff clay images of snakes are offered. These are temporary images similar to those used in the Nag Panchami ceremonies. Besides the "flag," Guga's whip is prominent. This consists of a ring, from which hang five iron-chain lashes to which are attached iron discs at intervals. Under special circumstances a bhagat lashes himself with two of these whips, one in each hand. The other instruments of worship are a trident upon which to hang the "whip," and a drum shaped like an hour-glass. But Guga is worshipped in the hope of securing other boons besides immunity from snake-bite. He is a powerful saint and so is worshipped in behalf of sickly children, and for help in a variety of diseases, and for the removal of the curse of barrenness.

Another saint more local in fame is *Gopāl Bāba,* who was an Aharwar Chamar, a shoemaker for a Raja. He was very badly treated by the Raja, who tried to kill him. The Chamar finally died from the effects of a nail which was driven into his foot. He is now a protective saint, and has his shrine under a nim tree in the village.

Still another saint of this type is *Devi Bāba,* who was a Dohar Chamar. His shrine consists of five chambers, one for one of his sons, one for another, one for his servant, one for Kalka and the central one for himself. An interesting thing about this case is that this man went to Bengal to learn magic. Upon his return he became famous by reason of the great number of evil spirits that he was able to bring under his own control.

A more widely known saint is *Hardaul, Harda Lāla,* or *Hardour Lāla,* now reckoned as a godling of cholera and of marriage. He was poisoned on suspicion of unlawful relations with his older brother's wife; but when it was discovered that he was innocent he was considered a martyr. He is now worshipped at weddings and during epidemics. A day or two before a wedding the women go to his shrine, worship, and invite him to be present at the ceremonies. His image on horseback is found on many shrines.

Still another saint of this type, *Dulhā Deo*, is the deified spirit of a bridegroom killed by lightning during the wedding festivities. He receives an offering of flowers in February and of a goat at marriage (the women share in the meat of this goat). During the wedding festivities he is worshipped in the cook-room, and oil and turmeric are offered to him. The ceremony is performed by the eldest son.

The less-noted saints are very numerous.

The next order of saints are those who have risen to the vague position of tutelary, or protective, godlings. Amongst these may be named *Bhīmsen* and *Bhishma*, heroes of ancient India. Bhīmsen is much changed from what he was as the famous character of the Mahābhārata, for he is now but one of the wardens of the household or of the village. His fetish is a piece of iron in a stone or a tree, or an unshapely stone covered with vermilion. With him there is found some pillar worship, and his giant strength is attested by some huge boulders in Kumaun, where his fingerprints are still pointed out. He is worshipped on Tuesday and Saturday, and offerings of he-goats, hogs, cocks, and cocoanuts are made to him. Bhishma is the childless one. Worship to him is performed in the month of February and in November-December, when lamps are sent to the houses of Brahmans. The housewife sleeps on the ground, on a place plastered with cow-dung. Lamps with red wicks and fed with sesame oil are kept burning in the house. Into each lamp a walnut, a lotus seed, and two copper coins are placed. Each evening during the festival the women prostrate themselves before the lamps and walk around them. They bathe each day before performing the ceremony. The bath is taken in the following way: Five lamps made of dough are placed, one at the entrance to the village, and the other four at cross-roads, under a pipal tree, in a temple to Śiva and at a pond. This last-named one is placed on a raft made of the leaves of sugar-cane. Grain is placed under each lamp. After the lights have gone out, the lampblack from the wicks is rubbed on the eyes and fingers of the worshippers, and the toe-nails are anointed with the oil that remains. During the period of worship one meal a day, consisting of sugar, sweet potatoes, ginger and other roots, is served. Flour is made from amaranth seed, millet and buckwheat. Butter is used, and only milk is drunk.

Further removed from the realm of sainthood, but probably connected with the worship of ghosts, are the tutelary godlings like *Ganesh* and *Hanumān*, both of whom are worshipped. Here both hero and animal worship are combined. Ganesh is the godling of good luck. Hanuman (Mahā Bīr), the guardian against demoniacal influences, is represented by rude images smeared with oil and red ocre. He is worshipped also as a cure for barrenness. He can assume any form at will. Because of his faithfulness to Rama he is the type of all fidelity.

The next class of godlings is those who are not, on the surface, connected with the belief in ghosts. However, these also are of human origin, and it is on this basis that they can be most easily understood. First, there is the preponderance of mother, or *śaktī*, worship. The local village demon-mother is universally feared. This phase of spirit-worship is connected with the worship of some form or other of *Kālī*, the consort of Śiva, and is without doubt of aboriginal origin. Sitala Mata, a form of Kali, has already been described. Besides her we have *Mātā Māī* and *Marī Māī*, or *Marī*. *Kalkā* and *Dakkanī*, fairly common in some sections of the country, are other forms of Kali. In some of her shrines there are three chambers and in these glass bracelets are sometimes found. Then there are the *Jungle Mother* and the *Birth Mother*. This latter goddess exercises powers which reside in a blue bead, Kailās Maura, which Chamaris carry to insure easy delivery in their practice of midwifery. Among other "Mothers" are *Bhūkī Mātā*, the goddess of famine, and the goddesses of the various crops. *Hulkī Māī* is a cholera goddess. Fickleness, and proneness to inflict diseases, unless propitiated by fealty, offerings and prayers, are all contained in this form of belief; and very many believe that Mothers have control over magical powers and over the secrets of nature. *Dhartī Mātā*, or *Dhartī Māī*, the supporter or upholder, is worshipped in the morning, at plowing time, at sowing time, and when a cow or a buffalo is bought. In the first milking after calving, and always at milking time, the first stream is offered to her. When medicine is taken a little is poured on the ground to her. With her is connected the belief in the sanctity of the earth.[1] With ceremonies connected with the worship of this goddess women are associated, and in some instances secrecy is practised. She also sleeps on the fifth, seventh, ninth, eleventh, twenty-first and twenty-fourth, or on the first, second, fifth, seventh, tenth, twenty-first and twenty-fourth of each month, so these and fifteen days of Kuar are sacred to her. On these days no plowing is done.

Another godling of the same type is *Bhūmiā* (sometimes *Bhūmiā Rānī*) of the homestead or soil, a protector of the fields. As *Bhūmiā Rānī* she is worshipped with cakes and sweetmeats, which are spread upon the ground in the sun and then eaten by the worshipper and his family.[2] Sometimes a shrine is erected to Bhumia when a new village is consecrated. His place is a domed roof or a platform. After harvest, at weddings, and when male children are born, vows have been made to Bhumia. Women take their children to his shrines on Sundays. Sometimes the first milk of a cow or of a buffalo is offered to him and especially after milk has spoiled. Young bulls are released in his honour. Seldom does he receive animal sacrifices; his are the fruits of the soil. When the crop is sown, a handful of grain is sprinkled over a stone, meant for his shrine, in order to protect the crop from hail, blight, and wild animals. At harvest-time the firstfruits are offered to him that he may protect the garnered grain from rats and insects.

[1] See the "magic earth" and similar ceremonies.
[2] See *Punjab Notes and Queries*, III. 56.

Bhairoṁ the terrible, sometimes called Bhairoṁ of the Club, or Bhairoṁ the Lord, is a form of Śiva. He rides a black horse, is accompanied by a black dog, and is, in the Punjab, a godling of the homestead. He is conciliated by feeding a black dog to satiation. He frightens away death. As the protector of the fields and of cattle he receives offerings of meat and sweets, and at his shrine spirits are poured out and drinking is indulged in. When one is very ill a vow is made to sacrifice a goat of one colour, and without blemish in case of recovery. In fulfilling the vow, the goat is taken by all the friends of the sick man to the shrine of Bhairoṁ, under a nim tree, just outside the town. Then the animal is beheaded with one stroke of a large knife, the head is placed before the image of Bhairoṁ, and a little liquor is sprinkled upon it. To the godling they say, "If you are pleased, let the goat's head open its mouth." The head always "speaks," if placed before the image with sufficent promptitude. Puris are then offered, after which the body of the goat and the cakes are taken by the sacrificers to their home to be eaten. The head of the goat is given to the gardener attached to the land where the shrine is located. At the house a feast is made, or if that be not possible, the flesh is divided among the friends and they take it to their homes. Bhairoṁ has *chelas,* or sadhus, called *bhopas.* When a man obtains a son, following a vow to Bhairoṁ, he dedicates the child to the godling for a certain term of years, and places him in the charge of a bhopa. Such dedicated persons are called bhopas.

Another village goddess, the godling of the village boundary, is *Chamuṇḍā,* a form of Kālī, one who delights in blood. On the outskirts of many villages there is a mound with some rude stones upon it to represent her.

There are several godlings of special interest. *Madain,*[1] the godling of wine, whom some call a demon, is greatly feared by the Chamars in the eastern parts of the Provinces and in Bihar. In Shahabad, for example,

[1] *Census Report, United Provinces,* 1891, pp. 220, 221.

Madain is the most serious form of oath taken amongst the Chamars, and a form of oath very rarely used, and then only when both parties involved are Chamars. They believe that whosoever swears falsely on this godling will suffer most severely. In the panchayats, when one man challenges another's testimony, he frequently calls upon him to swear by Madain. Sometimes, through fear, men, when so challenged, withdraw their testimony. The challenger has always to furnish the liquor, which his adversary then pours on the ground. The members of the panchayat are treated to drink by the challenger. If sickness or calamity follow, either to the man or to his family, it is attributed to his having sworn falsely. Chamars of Oudh hold him in great fear, but are ashamed to acknowledge him.

Saliya,[1] a special god of the Chamars, is worshipped with offerings of small pigs. Similar offerings are made to *Jakhaiyā* in fulfilment of vows when children are born. The pig is sacrificed by a sweeper, who marks the child's forehead with the blood.

Kale Gore Deo,[2] the black and white godlings, are worshipped daily by many Chamars, and by many other low-caste people. They are supposed to reside in a corner of the house where a pice has been buried, and are worshipped with offerings of food and drink. Their worshippers numbered, in 1891, about 750,000. Their origin is connected in some way with Kali Singh and Guga Pir, or with the two Mohammadan saints Kālu and Gori, said to be buried in the Partabgarh District.

Another localized divinity, *Purbī Deotā* (the Godling of the East Country), is worshipped at home. Once a year a pig, together with four yards of cloth, two loin-cloths, one nutmeg, four cloves and rice are offered to him. On the following day the worshippers bathe, after which they make an offering of bread and give a feast in which the flesh of the pig is cooked and eaten.

[1] *Census Report, United Provinces*, 1891, pp. 220, 221.
[2] *Census Report, United Provinces*, 1891, p. 220.

CHAPTER VII

THE MYSTERIOUS

THE Chamar attributes most events to spirit agencies, and, since he is always on the alert to outwit and to defeat these unseen powers, he is also always watching for signs that will indicate what is likely to happen, and he always plans to do things under the most favourable conditions. The whole field of luck and ill-luck and of omens is undoubtedly related to the belief in ghosts.

Like other folks the world over, the Chamar has his lucky and unlucky days. New clothes must not be put on on a Wednesday. To lend, or to borrow, on Saturday, Sunday or Tuesday is unlucky; and it is not wise to return from a journey on these days. Horses or cattle, or anything pertaining to them, such as leather, ghi or cow-dung, should not be bought or sold on Saturday or Sunday; and, if cattle should die on one of these days, they should be buried. The year's plowing should be begun on a Tuesday, a Wednesday, a Thursday or a Friday, or on the first or the eleventh of the month, and reaping should be begun on a Thursday and finished on a Wednesday. Cattle should rest on the 15th or the 30th of the month. The pressing of the sugar-cane should not be begun on Saturday or Wednesday.

Dis-á-sul is the demon of the four quarters. He lives in the east on Monday and Saturday, in the north on Tuesday and Wednesday, in the west on Friday and Sunday, in the south on Thursday. So it is not good to plow in those directions on these days. The south is unlucky and the cooking-floor should not face that way, neither should a person lie with his feet to the south.

Three and thirteen are unlucky numbers, and these days after death are very inauspicious. Odd numbers are generally lucky. Five and multiples of five, and compound numbers like four and a quarter, two and a half, and seven and a half, are lucky.

When one is starting on a journey, it is inauspicious to see a jackal cross the road from the right, a crow on a dead tree, or a dog shake his head so as to flap his ears. Under such circumstances a man should return home at once. The same precaution should be taken if a person hear an ass braying, or anyone sneeze near by, or if he should meet a washerman, an oilman, an eunuch, a widow, a water-carrier with an empty pitcher, a man suffering from disease or infirmity, a one-eyed man, or a man riding a buffalo. It is inauspicious if a brick drop out of the doorway, or if a cat is seen catching a rat, just as a person starts on an errand.

On the other hand, it is auspicious to meet a revenue collector (*lambardār*), a Brahman with his books, a man carrying a light, a sweeper with his basket full, a water-carrier with his pitcher full (one ought to drop a pice into it), or a woman carrying a male child. It is good to have a jackal cross your path from the left, to hear an ass braying on the left, or to meet a snake (it should be passed to the left and be greeted with "Salaam!"), or a loaded donkey (if he be loaded with clothes, but if loaded with bricks, unlucky), or to see a calf sucking its mother.

The following are good omens: To hear a jackal howling at night; to hear an owl hooting at night; to hear a partridge calling at night; to hear the voice of a koel in the morning; to see a monkey the first thing in the morning (but do not say "Bandar" before you eat); to meet the mantis (he should always be saluted when seen); to meet one carrying two full pots (*doghaṛ*) one above the other (they should be left to the right); to see in the morning a crow or a black buck. A man on horseback riding into a sugar-cane field during sowing brings good luck.

The following are ill omens: To see a pair of jackals in the morning; to meet a one-eyed oilman (very unlucky; unless he laughs he should be beaten); to see a cat crossing the road in the morning; to look on a barren woman the first thing in the morning; to see a dog flapping his ears or shaking his head when wōrk is in progress; to sit in or to sneeze into a winnowing-fan; and to have a kite settle on your house.

To see a cat or a crow throwing water on itself is a good omen. If a dog howl three distinct times at night a robbery is about to be committed, or trouble is imminent, or someone in the village will die. It is a bad sign for a dog with a bone in his mouth to come straight at a person. An owl hooting at night in a graveyard foretells death to the passer-by who hears it. Howling dogs portend evil because they are able to see evil spirits. If a spider falls upon a person it means that he will soon get new clothes, but the touch of a lizard is unfortunate. The owl, the kite and the cat are objects of dread, and the two former are bird-fiends of the lying-in room and of childhood. Up to the time of the performance of the chhatthi, the parents will not dry the child's clothes out of doors; and it is a very ill omen indeed if an owl, a kite, or a cat come into the birth-chamber. If, during the day, they put a child on a bed out of doors, they cover it with a sheet, and lay over this a piece of grass lengthwise of the body, that the shadow of a passing kite may not fall upon the child and cause what is known as *chīlwās*. If a man reach a village at dusk, or after nightfall, and hear a woman crying, he must go back home at once, or at least go as far as another village to rest for the night and then go home; or, he may sit down and smoke and go on. If a newly purchased horse, on seeing his owner, shakes his head, the bargain should be broken off; but if he paws, it is a good sign. A one-eyed man coming to a party stops the merriment (he should be driven away). Persons with defective eyes are constitutionally vicious and cunning (*chālāk*) and they should be avoided. A person who dreams himself dead will live long; but one who dreams that he is well dressed, or that he is going out, will die soon. The itching of the right palm signifies wealth; of the left, indicates that money will be paid out. The twitching of the upper right eyelid signifies good; of the left, ill. The indications in the twitching of the lower lids are the reverse of the above. The twitching of the left eyelid of a woman signifies joy. Sneezing is lucky, and as a man cannot die for some time afterwards, he should be congratulated.

Another mystery is that which is included under the general term of evil eye, or fascination (*nazar*). It is believed that there results from the look or glance of many persons, and it is by no means certain by how many, sickness, wasting diseases in children, death, misfortune, loss, calamity and diabolical influences generally. Such glances affect individuals, cattle, crops, food, houses and almost every other thing of value, including building enterprises and handwork. The real root of the matter is in covetousness. Any especially beautiful or perfect person, animal, or object is subject to fascination, and the influence is due to a desire on the part of some person or spirit for the particular thing or quality found in the object, but which is lacking in the spirit or person exercising the influence. One-eyed persons, those born during the Solono[1] festival, black-tongued individuals[2], persons who have had no children or whose children have died, and those who have eaten ordure in childhood have the power of fascination. On the other hand, evil-looking and deformed persons (as those born with double thumbs or fingers or toes, the blind, and the lame), bald persons, those born during the Solono festival, and those who have eaten ordure or cow-dung are not subject to the influence of the evil eye. Likewise defective or spotted or imperfect things are not subject to this influence.

[1] The Solono, held in August, is connected with the worship of Rāja Bāshak. It is not a mela but a domestic festival. It is at this time that vermicelli is made and offered with ghi in the home-fire. There is a feast at this time. Only those who worship Rāja Bāshak make the vermicelli, and it is made but once a year.

[2] Black-tongued individuals seem to be those with pigmented tongues, those whose abuse and prophesies always come true.

Fascination is a potent power, and since no one ever knows just who many exercise this power, it behoves most people to do everything in their power to avoid it. The power is exercised through a person's eyes. The explanation is that evil spirits are always on the look-out for opportunities to exercise their untoward powers, and that they often take their cue from the way in which people look at persons or things. The belief, then, is explained by reference to the world of unseen spirits. So, nearly, if not quite all, of the devices which are effective in scaring ghosts are potent in this field also. Such means as are used have for their purpose the catching of, or the diverting of, the evil glance before it actually rests upon the object for which protection is sought. Movable property is marked with black spots; charcoal is placed in the basket of food that is sent to the field; an earthen pot is blackened and placed in the thatch, or carelessly left in the courtyard, or hung on a pole in the field; and sometimes bits of blackened rags are stuck into the mud-wall while it is in process of construction. Blemishes are left in things of value, and other devices are used to elicit the feeling of disappointment or of disgust in the minds of those who might otherwise look upon them with favour or satisfaction. So, a bit of food that has been bitten into is sometimes put into the basket of provisions that is sent to the field; and the head of an animal is set on a stick in the crops. Houses are protected by putting an old shoe, heel upwards, in the thatch; by driving nails into the doorposts and the threshold, and by fastening horseshoes in the threshold; by hanging up a *baya's* nest (Indian weaver-bird); by fastening a hedgehog's skin or porcupine's quills in the doorway; by making marks of various kinds on house-walls; by drawing a picture of a churel in a conspicuous place, especially on a good house; and by throwing mustard-seeds into the fire. Horses and ponies are protected by beads hung around their necks and by leather covered with gold leaf, shaped as a single or a double triangle. Grooms often weave into the horse's tail a clashing-coloured string or rag, often with a kauri and

POTS USED AS OFFERINGS

POT SET UP AS A PROTECTION AGAINST THE EVIL EYE

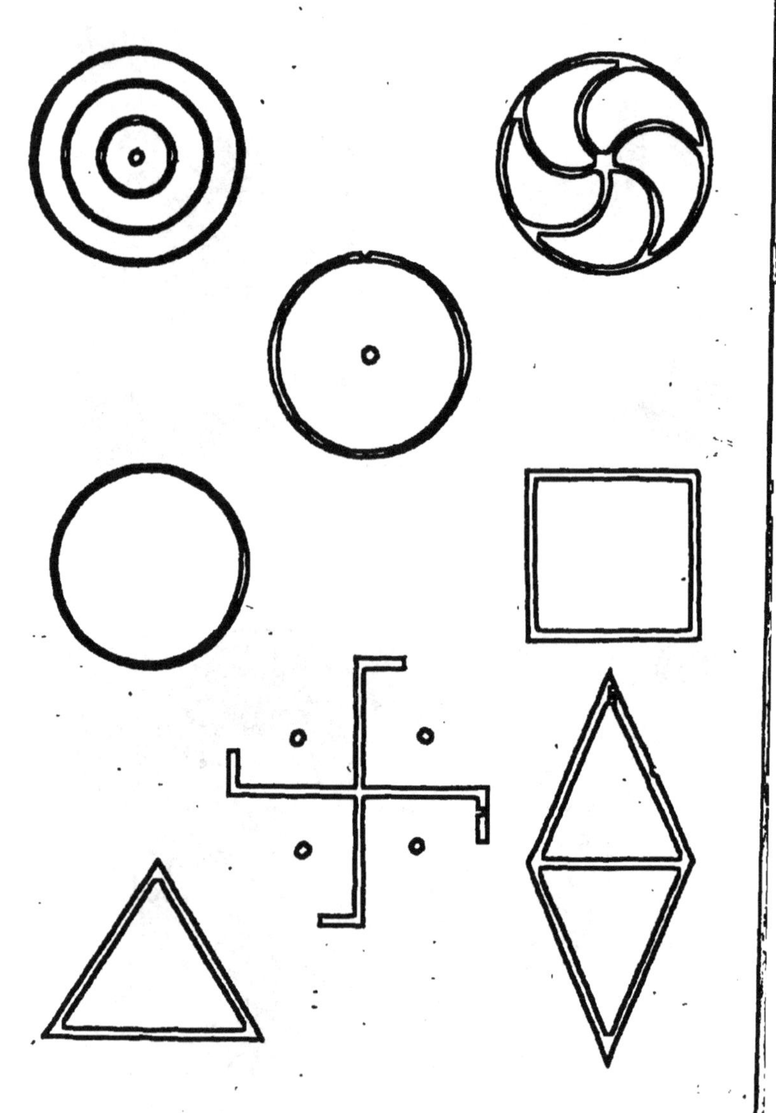

PENCIL DRAWINGS OF MAGIC SYMBOLS

two triangles of broad cloth, one red and the other black, fastened into it; and sometimes blue-black threads are tied around the fetlocks. Often, when a horse is eating, a duster is thrown over his withers. Cattle are often protected by a bit of turtle-shell or skull and an iron ring tied around the neck with catgut. Kauriaṁ and blue beads are used in the same way. A blue rag is made into knots and is then hung around the animal's neck. Sometimes such articles are tied around the base of the horns.

The greatest care is exercised to protect children. Since a glance which results in complete satisfaction to the one who casts it is the serious thing, efforts are made to create disgust by making ugly and repulsive objects most conspicuous on attractive persons. Black, as the colour of mourning, is not attractive, so it is used. Lamp-black is rubbed on the eyelids and the forehead; the face and the teeth are blackened. Sometimes a child is dressed in filthy clothes, or is left unwashed for years, or opprobrious names are used, to create the attitude of disgust. Thus a mother who has lost a child by smallpox will name her next child *Kūṛiyā* (he of the dunghill). Again, the mother will attempt to save her child from evil glances by giving him a name that will indicate that he is of very little value. So children receive such names as Pāñch Kauṛī (worth but five shells), and Māṁgtā (begged, a mere gift). Further, names of devil-scarers are given to children, and those include the names of godlings and saints. Deceit is practised by dressing a boy as a girl, for thus the evil glance is robbed of its power.

Some devices to counteract the apparent effects of the evil eye are interesting. Most infantile troubles are attributed to the evil eye. If a baby (or a calf) is restless and will not take its food, relief is obtained by grinding three red chillis and waving them over the sufferer three times, and by then throwing the offering into the fire. Another effective remedy for infantile troubles is the following: Take bran, powdered chillis, salt, mustard and the eyelashes of the child; wave them seven times around the child and throw the whole collection into the fire. The bad smell that arises signifies relief. To protect a child who has been carried on a visit to another village, seven little stones should be waved each seven times around the child's head and thrown in seven directions before the child is brought into its own home.

Many devices, such as jewels, glass beads (especially blue ones), mirrors, and other bright objects, amulets, charms, flowers, and turtle-shells are worn. Blue tattoo marks, especially between the eyes or over the eyebrows, are effective protectors; and ear and nose borings are resorted to, and, in case of a long-wished-for son, these are made soon after he is born.

Instances like the following illustrate how prominen the belief in fascination is: A Chamar woman, wh child was stillborn, thinking that her misfortune was to the influence of the evil eye, wrapped up a piec cloth, used at her confinement, with two leaves of b some cloves and a piece of castor-oil plant, and thre down a well. This was done, probably, under the adv of a wizard. Similar articles are sometimes burie cross-roads.

The use of lampblack by women is gracious, since one who puts it on her eyes cannot cast the evil eye.

Regular charms against the evil eye, called the *na-bantā*, may be bought in the bazaars.

The description of the means employed to avert the ev eye could be much extended, but a reference to th enumerations under the topic of devil-scarers will suffice.

The cult of the mysterious occupies a large place wit the Chamar. Magic, which has the spirit-world for it background, has to do with the control of life and destiny. The primitive man makes no clear connection between cause and effect, but associates together all sorts of things which to the modern man are unrelated. And, since the cult of magic grows out of confused notions regarding the cause and sequence of phenomena, primitive peoples seek to accomplish ends by means that civilization recognizes at wholly inadequate. There can be no successful art of witchcraft, however, until the possible are distinguished from the impossible means, that is, until the magician has a clearer notion of which sequences are possible than his clients have. Magic may be directed to public or to private ends. It is collective and ceremonial as well as individual and secret. The former is usually for the public weal, and is socially approved; the latter is almost always for nefarious ends, and is more often tabu. It is witchcraft. "The occult is marked by divergence in trend and belief from the recognized standards and achievements of human thought."[1] Magic may be classified as sympathetic and imitative. The latter, which is also called symbolic magic, is based upon real or supposed resemblances between things. The former is based upon material connection between objects, and proceeds upon the conviction that, whatever is done to a part or possession of a man, is done to him. Probably four-fifths of mankind believe in sympathetic magic.

Magic, in all its primary elements, is found in the practices and beliefs of the Chamar, who, in the nefarious branches of magic, commonly called the black art, bears an unsavory reputation. In the following pages private and public magic will be discussed in turn, taking first the sympathetic and then the symbolic phases of each branch of the subject.

There is, first of all, the belief that objects which have once been in contact with each other are still effectively related, even though they may be separated, and that, whatever may be done to one of the objects, similarly affects the other. Thus, control may be obtained over a person by getting possession of his nail-parings, his hair, his blood, his saliva, or something connected with him.[2] For this reason the cord and placenta are disposed of so carefully after parturition. Another illustration is that where a Chamar woman, who had lost her children one byone in infancy, came secretly to the healthy children of a Christian and nursed them, that her weakness might be transferred to them, and that she might get possession of the power that would result in healthy children. Another

[1] Jastrow, *Fact and Fable in Psychology*, p. 3.
[2] Such practices are almost always under the direction of a magician.

Chamari, whose children were all girls and who was anxious for a son, took some of the blood of her confinement and sprinkled it upon the clothes of a boy. In this connection other examples of magic which has for its purpose the obtaining of offspring may be noticed. A childless woman will surreptitiously cut a corner from the chadar of a mother who has a large family; or she will steal the clothes of a child belonging to a large family. For the same end little children are sometimes fed by barren women. Another illustration of sympathetic magic is that where the pipal tree is worshipped with the hope of obtaining offspring. Cotton thread, made by a virgin, is taken by a barren woman to a pipal tree. There she offers batasas, water and flowers, and then walks around the tree seven times, winding the thread about the tree as she goes. The offerings, together with the thread that remains, are then left at the foot of the tree. The woman then prays, and makes a vow that if children are given to her she will hang a flag in the top of the tree. In this ceremony material connection is secured with an object (the tree) which is full of vitality. The use of mango-leaves, betel-nuts, walnuts, almonds, cocoanuts, and plantains in ceremonies related to fertility is because these are all products of fruitful trees. Monkeys, ants, the black buck, and the peacock are worshipped, probably because they are so prolific. Rags of clothing are tied on trees, nails are driven into trees, and pins are thrown into wells, in order to get into ceremonial connection with the spirits identified with these respective objects, and thus to be assured of the benefits which they are able to bestow. When an elephant passes through a village, its footprints are touched, especially by children, in order that they may secure the benefits of its great strength, and elephant hairs are worn that a person may be able to overcome the power of fascination.

Various things are eaten with definite purposes in view. Crow's tongue is given to children to insure long life or to help them to talk. A weak child, or one subject to pneumonia, is given small bits of tiger's flesh to eat. The heart and liver of the Indian badger are eaten to scare away demons, because the animal digs into graves and feeds upon the dead. The eyes of an owl are eaten for the gift of wisdom and that a person may be able to see in the dark.[1]

In carrying out the same practice of sympathetic magic certain things are worn. Amongst these may be mentioned the flesh of the tiger (and that of the leopard and panther), the claws of the badger, the horn of the jackal, bear's hairs, and those taken from a horse's tail. Besides these the long betel-nut is used; and articles such as shells, turtle bone, boar's tusks and teeth, snake bones, the leather of an old shoe, turquoise and knotted threads are worn, and rings made of a combination of metals are used. In many instances some of these articles form part of the contents of a charm or amulet. Immediately after birth, parents often hang about the neck of their child an amulet, containing, amongst other things, tiger's whiskers. In amulets many articles used as devil-scarers may be found. Another class of objects which are considered as possessing powers in themselves, and as conferring good luck upon their possessers, are called talismans. Both the amulet and the talisman may be combined in a charm. Bags made of leather or of black cloth, or small cases of metal (silver, copper or brass) are hung around the neck or fastened on the arm. In them are found an odd variety of articles chosen because of their magical significance, and there may be besides, in the charm, a bit of paper upon which is drawn a square representing a small seat with the figure of some godling, as Hanuman; or, upon the paper may be written some mantra; or some mystic or some clear symbol or impression may be drawn upon the paper. The cases and bags are usually of some simple geometric shape. A characteristic amulet contains pictures of the godlings Mahabir and Bajrang, a bit of paper with a mantra written upon it, two and a half grains of rice, two and a half grains of urd, two and a half grains of barley, two and a half cloves, and a bit of a parasite (*amar bel*). (Some say that Chamars

[1] See under folk remedies.

who eat beef may not wear such amulets.) Pouches made of leather and consisting of two parts, in which are put chillis, mustard seed, salt, husks of barley, charcoal and haldi, are hung around the necks of cattle to protect them from the evil eye. An interesting charm is that which the second wife wears about her neck, a little metal amulet upon which is scratched a representation of the first wife. This is called the Saut Sāl, or Saukan Maurā. All the marriage presents made to the second wife must be offered first to this charm. When she puts on fresh clothes or jewelry, she touches them first with the image as a sign they have been offered to the spirit of her predecessor. If this is not done, it is believed that the offended spirit of the first wife will bring disease or death. If the second wife, or the husband, die soon after the marriage, the death is attributed to the first wife, who has not been suitably propitiated.

The use of magic for nefarious or ulterior ends, both in the making of love-charms and in working injury to others, is common. In almost all cases these devices are used under the advice and with the help of a witch or wizard. Since owl's flesh makes a person a fool, wives often give it to their husbands, in order that flirtations with other men may be carried on without arousing suspicion. Charms are repeated over mustard seeds, which are then placed in the path where the person towards whom the incantation is directed will walk. If he step over the seeds he will be sure to be under the spell of the one who employed the wizard. Thread is often used instead of mustard seed. At other times a knotted thread or the charm is buried at the door of the person to be brought under the love-spell. If he or she step over the buried object the charm will be successful. Occasionally the person who wishes to bring another under his power wears a specially prepared charm on his arm.

The following procedure is a good illustration of the use of symbolic magic: At night a person goes to the Ganges, strips off his clothes, and proceeds to cook rice in a skull. While the rice is being prepared evil spirits gather about it; consequently a good deal of courage is required on the part of the one who is carrying out the magic rites. He must refuse to listen to noises or to be frightened. When the rice is done, he must go with a rush and throw it against a tree. The rice which adheres to the tree and that which falls to the ground are collected separately. The rice which stuck to the tree becomes a potent love-charm upon anyone to whom it may be given to eat (of course, secretly). That which fell upon the ground may be used to break the charm, or as a means of working evil upon the one to whom it is given.

A darker chapter is opened when we come to the magic which is used for injurious or destructive purposes. No doubt, before the strict hand of the present Government was laid upon violent actions, magic had some gruesome tales to tell. There is now no trace of the old custom (symbolic magic) according to which a barren woman, or one who had lost a child, caused the child of someone else to be murdered, bathed in its blood, or drank it, that she might bring the spirit of the murdered child into her own womb. Envious women will take ashes from the burning-ghat and cast them over children, believing that the children will waste away under the spell of Masani. Thieves throw such ashes over houses, to cause the inmates to sleep soundly during the burglary. Injury is done to a man by doing to his shadow what the person wishes to do to the man himself. In order to kill an enemy, a man will divide a pumpkin, a cucumber, or a lemon, believing that the enemy will thus be made to suffer. Of course, these devices are carried out by means of spells and incantations and through the agency of spirits under a wizard's control. Another nefarious practice is to place a goat's head and liver and a knife in an earthen pot, pronounce a formula over it, and send it to an enemy's house. When the pot falls upon the roof blood will gush from the mouth of the victim and he will die. By means of charms, an earthen pot (*hāṁḍī*), in which certain things and a lamp are placed, is sent to a man who is to be murdered. The pot flies through the air, falls upon the place where the victim is, and kills him. Care must be taken, before the pot is set in motion, that a sacrifice be offered for each stream that must be passed over, otherwise the pot will fall into one of the streams and the magic will fail. Some make an image of the enemy out of flour and place it with a four-wicked lamp on a tray. Before this a fire-sacrifice is performed and vows are made. Still others drive knives or needles into an image, and cast it into a tank or a stream or bury it. And some cut the throat of an image or put out its eyes. All of these devices are for ulterior ends. Sometimes a disease-godling is sent to prostrate an enemy with dysentery or some other serious malady. For example, a Chamar was sleeping in his bed which was set up over the crops in a field. In the night he saw Bhawānī (a woman, a *devi*) coming towards him. The man took her to be his master's wife, so he stood up, saluted her, and asked her why she had come. She did not make answer, and the man got frightened. When he started to get down she seized him by his private parts and injured him very seriously. A wizard was called in, who brought the man around by means of magic. The goddess had been sent by an enemy. In curing the disease the wise man used the following charm addressed to Mari: "I call upon the name of Mari. O Mari, be thou my deliverer. The dead of yesterday, fallen into the depths, Mari goes to unearth. May the birds (*e.g.* Birs) of the different countries be enchanted, but the wells be unaffected by magic (*e.g.*, May the women as they come to the wells be brought under the spell, but not the wells). May those who sit in order (*e.g.* in the feast or in the festival) be brought under the spell. May thine eyes be brought under the power (of my) enchantment. Let me see thy magic power and the strength of thy spells (*e.g.*, O Mari, exert thy powers in my behalf)."[1]

To bring death upon a person, a man will give an owl liquor to drink, and for forty consecutive nights will pro-

[1] *Marī, marī maini Karuṁ, Marī hoe sahāī.*
Bāsī murdā gau pātāl, Marī ukhāṛan jae.
Des des ki chirya phaṁse kuāṁ na phaṁse.
Panghat phaṁse, phaṁse tere naina.
Dekhuṁ teri sakat, tere mantar ki sakat.

nounce his enemy's name to the bird. If the owl has been kept so carefully that it has heard no other name during the whole period, the magic will be successful. The name is an integral part of a person, and, therefore, to know a man's name is often to put him into a wizard's power. In such cases the magic is performed "upon the name" of an enemy. The spirit or godling by whom the spell is being cast must not speak to anyone while on its mission. To escape the danger of the black art, children frequently receive two names, one of which is kept secret.

Besides the belief in the power of magic spells, pass-words, and names uttered by the authorized persons in the prescribed manner, there is a magical significance given to certain tabus. The wife may not use her husband's name, nor the name of her father-in-law (often), nor that of her mother-in-law. The wife of a younger brother may not use the names of her husband's older brothers, nor touch them, nor speak nor appear before them unveiled. But conditions are such that it is impossible strictly to observe these rules regarding the use of the chadar. Some persons may have neither doors nor door-casings. To erect these would mean a death in the family. To avoid this calamity, if they desire a door they build it while the dead body of an old man lies in the house. Certain foods are tabu on penalty of snake-bite; and the wizard may forbid certain foods to certain persons. Some may not wear black bracelets. Others may not sow pumpkin or gourd seeds; still others may not eat washed dal. Some must live in *chappars* (a thatched-roof hut); others will neither make rope not string charpais in the month of July. (This tabu is evidently connected with the prevention of snake-bite.)

Public magic is mostly communal and for ends acceptable to the community. It is largely of the symbolic or imitative type, although there are some elements of sympathetic magic involved. Rain is essential to the securing of the harvest, and two sorts of devices are employed to secure rain. One is directed against the magic of the Banias, who try to prevent rain; and the other aims to secure rain. When clouds are overhead,

Banias and shopkeepers fill their lamp-saucers (chirags) with liquid ghi. Ghi is one of the products of the cow, and so tends to scare away the storm-demons.

To bring rain a plow is hung in a well. Sometimes a chapati and a new earthen pot filled with water are placed upon five clods in a field. If the water in the pot causes the bread to mould, rain will fall. Rice and sugar are placed where four roads meet, and sometimes the road-crossing is defiled. A number of women take a four-year-old girl into a courtyard of a mahalla; there the women sing, dress her in new clothes, and ask, "Will it rain?" She answers, "Yes"; and they ask, "When?" Ten or twelve boys go from house to house, lie in mud and repeat the verse: "The time for rain has come; let the rain fall." As they go they take up a collection, with which ghi, sugar, flour and vegetables are purchased and cooked near a tank. The food is offered to Brahmans, to the cow and to Indra, and then eaten. An elaboration of this device is as follows:

When the rains are delayed an interesting ceremony is performed. The faces of ten or twelve boys in the village, or mahalla, are blackened. Every boy takes a stick in each hand, and then visits every home in the place. At each house they call for water, which they throw on the ground, and then they proceed to wallow in it. They beg also, and, as they go from place to place, they strike their sticks together, crying:

> "Heavy showers will fall in front of your house:
> It will rain; the goddess will send rain;
> The paddy shall be sown;
> The old men shall drink the water of the boiled rice,
> And the young men shall have rice to eat."

Of the proceeds of the begging, grain and flour, the former is ground and then mixed with the latter. The boys then repair to a tank, or a river-bank, where a space is cleared and a fire lighted. The fire is removed, and dhak-leaves all spread, upon which the cakes made from the flour are laid. The cakes are then covered with dhak-leaves, and fire is laid over the whole, and the cakes are well baked. They are then offered to Indra and eaten.

In this we have the imitation of the black rainclouds, and the fiction of floods of water combined with the worship of Indra, the thunderer, who defeats the demons and liberates the water. This is a dramatization of the storm. A person should not rush out of the house bareheaded during a shower, unless he wishes the rain to cease. Rain may be driven off by the following device: Take a pint and a quarter of rainwater, put it into an earthen pitcher (ghara) and bury it in the ground at a spot where a roof-spout discharges.[1]

To prevent a hailstorm a tawa or a sieve (sup) is pounded with a bamboo. In this we have the use of powerful means to frighten away devils. The demon is opposed by iron, by black colour and by noise, or by the sieve-fetish. Probably the sound satisfies him that hail has already fallen. Sometimes over urd kí dāl a magician repeats mantras and then throws it on the hailstones. In this way the storm is confined to a single field. The village recompenses the owner of the field.

There are numerous practices connected with sowing, seedtime and harvest. During the sowing of *bājrā* (a species of millet: it is *par excellence* the "food" of the poorer classes) or *jawār* a request for fire should not be granted, for, if coal is given, the blades will be eaten by worms and so will turn black as soon as they spring up. At the sowing of the sugar-cane, the person who proposes to sow calls his labourers, or friends who are going to help him, saying, "Come, I am going to sow sugar-cane." Choosing a boy who has not been married, and whom they call "the bridegroom," they dress him in fine clothes and set him on the boundary of the field. No one would enter the field with his shoes on. They take one cake of gur with them, which they offer in the name of Bhimsen. They rub some on the plow, and the remainder they distribute amongst themselves. Bhimsen, thus propitiated, is sure to protect their crops. They then begin to sow. A boy is chosen to follow the sower and cast any stray joints into the furrow. He is called "the crow." The sower

[1] *Punjab Notes and Queries*, III. p. 115.

dare not look back. While this ceremony is being carried out in the field, the women at home are preparing a special dinner. When the food is ready, they draw marks on each side of the door with cow-dung, to imitate a tall, abundant harvest, and then take the dinner to the field for the workers. For the feast of the sugar-cane planting only such foods are cooked as will properly mix, for example, khir and karhi. A dinner of *khichṛī* would be inauspicious, for the dal and rice remain separate, and the sugar-cane would come up scattering, and the harvest would be light. Before the sowing of wheat five small earthen pots are filled with the grain, placed upon the ground and trodden down. Sometimes men throw a little grain in the direction of the Ganges, or they bury a fistful of grain in each of five places on one side of the field. Before each of these places five men face the Ganges. After the sowing is completed a feast is held and a fire-sacrifice is offered before a plow. The owner of the field must not be hungry at the time of the sowing; he must do his sowing after a full meal. To get rid of weeds that are persistent, bury in the field the placenta from the first calving of a cow.

Before the cutting of the grain, offerings of firstfruits are made to various godlings. Bhairoṁ receives such offerings at the Holi in the house-worship. Sometimes a little grain is left standing in the harvest-field. After a short rest the reapers attack this last bit, tear it up and throw it into the air amidst shouts of victory to their local godlings. Sometimes this last bit of standing grain is left uncut for the spirit of the field. At other times it is attacked by women, cut, mixed with other grains, taken home and boiled. This preparation is then passed around and eaten. Often some cash from the first sales of the grain are brought by the Chamar to the landlord, and it is believed that the Chamar brings luck with the money.

At the beginning of the sugar-cane harvest what is known as the *Gayās* is celebrated. The harvesters, going into the fields, bind a few stalks of the cane together at the top. On the ground, beneath the knot, a small pot is placed. This vessel is quickly filled with water, that the vat beneath the cane-press may be abundantly filled with sap. A fire-sacrifice is then performed, and the workmen go around the field one, three or five times. After this they break off and bring home some of the sugar-cane. These first stalks are offered on an altar, called *makhas*, or they are placed on a cot together with an axe, a shovel and a sickle, and over these a piece of new cloth is thrown. A fire-sacrifice follows. Then women come and sing the praises of the godlings. The cane is then cut into pieces and distributed. The first stalks are cut by the man who is to collect the juice for boiling. While the ceremonies are going on in the fields the women at home are busy cooking rice and urd for a feast. On the walls of the house a figure is drawn, and a fire is lighted, and a basket is waved over it, while the worshippers repeat: "*Uṭh Nārāyan, baith Nārāyan; chal chhane ke khet men maiṁ kāṭuṁ; tū chhet Nārāyan.*" Some of the food is then eaten and some is distributed to beggars and friends. The few stalks that were bound together at the beginning of the harvest will not be cut until all the rest of the field has been harvested. At other times the first few stalks that are cut are taken home and crushed, and some of the juice is offered to Chamunda, some is poured out on the ground, and then the remainder is boiled down. The gur thus made is distributed amongst the men, and some is given to their sisters and daughters.

At the beginning of the cotton harvest, an especially large stalk is chosen, and balls of cotton are fastened to it. When women begin to pick the cotton, they go round the field eating milk-rice, the first mouthful of which they spit out on the field towards the west. The first that is picked is exchanged at the village shop for its weight in salt, which is prayed over and then kept in the house until the picking is over.[1] At other times a fire-sacrifice is performed in the middle of the field, and women eat rice there. When they come home they throw a little of the cotton on the thatch.

[1] *Report, Census of Punjab*, 1881, p. 119.

The threshing and measuring of the grain is important. The men go to the threshing-floor carrying articles for worship such as milk, ghi, turmeric, boiled wheat and a variety of grains. The threshing-stake is washed. Boiled wheat is scattered about in the hope that the bhut will be content with that, and that he will not require any of the new grain. Before winnowing the grain, two double handfuls are taken out and one is given to a Brahman, the other to Bhairoṁ. This is done to keep away demons. Then five baskets of the threshed wheat are winnowed, and the chaff and the wheat are measured separately. If the baskets turn out full, or if there is an excess, it is considered auspicious. If the measure be short, another place is chosen for the winnowing. The winnowing follows. Then the grain is heaped up in one place and a khurpa, a broom, dub-grass, and cow-dung are brought for a ceremony, and an incantation is repeated. When the winnowed grain is heaped upon the threshing-floor an incantation is pronounced:

> "Lord God of the corn-heap, give a hundred blessings.
> Corn-God and Lord, multiply a thousandfold.
> God, give us prosperity in our affairs."[1]

Every night the grain that has been winnowed during the day is measured, the company keeping perfectly quiet. The number of baskets is recorded by knots or by grains. Spirits steal the grain that is not measured. The first scale-pan of grain is taken home, and part of it is given to Brahmans and gurus, and the remainder is made into bread and distributed along with sugar.

In the grain-bin, which has a hole in the side near the bottom for taking out the grain, a sun symbol, for protection, consisting of a circle with covered radii, is sometimes drawn.

Another illustration of public magic may be added in conclusion. When a well is to be dug, little bowls of water are set out around the proposed site on a Saturday night. The one which dries up last marks the exact spot where the well should be dug. They begin to dig leaving

[1] *Punjab Notes and Queries*, Vol. I, p. 40.

the bowl intact, and the clod on which it rests is the last to be removed. This clod is called Khwaja Jī. The saint by this name is worshipped and Brahmans are fed. If the clod should break during the digging of the well it is a bad omen, and a new site has to be chosen a week later.

The use of folk remedies and the practice of primitive medicine runs into the world of the mysterious, into the realms of magic, and into the practice of the wizard's art. To guard against sickness, oil distilled from owl's flesh, or from that of the flying-fox is rubbed on the body; turtle's flesh is a tonic which cures indigestion and prevents rheumatism; lizard's tail is a cure for fainting-fits; crow's tongue cures stammering and dumbness; the badger's liver and its entrails, ground and mixed with milk, are used as a cure for diarrhoea; rabbit's flesh is a medicine, and its entrails, powdered and mixed with mother's milk, are a cure for pneumonia in infants; the hoopoe's flesh prevents heart disease; the oil distilled from the flesh of the *dhanesh* (a bird with a very long bill) is used to relieve rheumatism; pills made by mixing spider's web with gur and bedbugs are prescribed for fever and ague; the fluid fried out from the cooking of a goat's liver is put into the eyes on a Sunday or on a Tuesday to cure night blindness; hare's blood mixed with mother's milk and likewise pea-fowl's legs are used in the cure of fever; to cure deafness a peacock is boiled in oil, and the compound so secured is dropped into the ear; the contents of the kite's eye is used to treat weak eyes; bat's bones are tied around the ankle to relieve severe pain in the bones of the legs; rabbit's blood is used as a cure for white leprosy; peacock feathers are placed on the bed to prevent nightmare; a porcupine's intestines are dried, powdered, and given in water to cure wasting diseases in children; the juice of onions is used to cure blindness; human urine is prescribed for whooping-cough; copper rings are worn to keep off pimples; and a hen is fed to relieve a child's trouble in teething, or the child's mother's brother takes him on his hip and plows the fields.

An interesting protective ceremony is the following: Near the entrance to a courtyard, or at the corner of the house, water-jars are placed. These are sometimes explained as placed there so that a devil or ancestors may come there to drink. The offering of these water-jars is as follows: Before four jars placed in twos, one above the other, a fire is lighted in an oil-lamp. Into this fire oil is sprinkled by an old woman, so that the fire is kept burning brightly. A batasa is placed before the fire. The women utter prayers to the local godling, some form of fire, and, gathering around, sing and laugh and chatter. Then, one by one, they come forward and bow to the ground before the fire. If they have babies with them they touch their heads to the ground, holding the children by the feet head-down to do so. A young woman then lifts a pair of waterpots to her shoulder and then to her head, and, followed by the group, walks through the courtyard and around it, and then to the place where the waterpots are to be placed. Then the other pair of pots is placed in a similar manner. The group of women then return and partake of sweets. The whole object is to ward off disease, in some instances to preserve the grandson, where the husband and sons are dead. The godling propitiated is not malignant, but she may save them from sickness.

Snake-bite is often attributed to the anger of snakes, and so, when a man is bitten, he calls upon the snake, in the name of Zahra Pir, to forgive him. A cure may be wrought under the direction of a medicine man. The person who has been bitten must sit on the ground with a sheet thrown over him, so as to completely hide his face. If the patient be a woman she must unloose her hair. If the person be unconscious, a twig of the nim tree must be placed on his head, or he is set on nim-leaves. If he be in his senses, ghi and black chillis are administered to him. One or two large covered earthen vessels are placed near him. The lids of the vessels and cymbals are beaten, or a metal tray is placed over a large earthen vessel and beaten with a stick, and songs are addressed to Raja Parikshit. This Raja was once bitten by a snake. These songs are sung to please the king of snakes and also to please the snake which has bitten the man. Then the spirit of the snake comes upon the man, who straightway begins to dance or to wag his head. Whatever the patient speaks while in this state of frenzy is spoken by the snake-spirit, and so any orders given are carried out immediately. For example, he directs them to sing a certain song a certain number of times. When the order has been carried out the snake leaves the man, *i.e.*, the patient is cured. When the snake-spirit comes upon the man it explains to him the reason for the calamity, and tells him what offering must be offered to secure recovery. These offerings will be, usually, a goat, bread, or clothes, which of course go to the wizard. Sometimes Zahra Pir himself comes. In that case they intercede with the saint for the man's recovery.

The man must be kept standing; if he falls down he will die. Some think that each snake is accompained by an evil spirit, which enters the body with the bite, and that this spirit must be exorcised to save the man. So the spirit must be driven out by a more powerful demon, or by means of some powerful devil-scarer. This is exorcism. Nona Chamari is worshipped, in the hope of being saved from death from snake-bite. Those who die of snake-bite are buried. Since it is believed that the person who was bitten lives on for six months, the body is not burned. The hope is that somebody expert in curing snake-bitten people may come along and discover that the person buried there had been bitten by a snake. Such a person would have the grave opened, and then, by the use of spells, would resuscitate the man. The body of one dying of snake-poison is sometimes thrown into a stream, with the hope that it will float along until, by some chance, it comes under the influence of one who might restore the dead to life. Salt is sometimes put into the eyes of one who has been bitten by a scorpion.

In case of malarial fever, offerings of a concoction of milk, hemp-leaves and sweetmeats are made. A service takes place on a dunghill. A black pot is whitened, marked with haldi, then broken, and the fragments are carried away by children. To prevent fever five batasas, or gur, are waved around the person and then cast into a river. Or water in which cloves are thrown is waved and then poured out at the foot of a nim tree. Bits of gourd are waved about the head of the patient and left at cross-roads. Another practice is to go quickly to a tank when the fever comes on and to remain there until the chill begins. The person then bathes and remains sitting in the water for two are three hours. The spirit causing the fever will not come into the water. Rags are tied on trees, and water is poured out at the foot of a tree, to obtain deliverance from the fever-demon.

Another method employed to cure or to prevent disease is to propitiate the demon responsible for the trouble. Sometimes a bhagat is called in, who identifies the spirit of the disease, and explains what must be done and what offerings must be made. To identify the demon causing a disease the wizard hangs a scale-pan from his forehead with his hand. Into this pan tobacco, furnished by the sick person, is placed. The wizard then begins to name over demons slowly. The name that is being called when the scale-pan begins to swing is the one who is causing the trouble. In some instances a huqqa is used in place of the scale-pan to identify the disease-demon. The wizard puffs away slowly, naming demons of disease. The name on his lips when his head begins to jerk indicates the demon.

Sometimes Burha Baba's help is enlisted. In this case someone goes to a tank where this saint lives, makes vows and offerings, and brings home some mud, which is placed on the sufferer. This saint is entreated for those suffering from ringworm, and especially for those attacked by white leprosy. He receives offerings of pigs' ears and the flowers of the *Chāṁdnī* plant.

Many devices are employed to coerce the spirits of disease and relieve the sufferer. As already suggested, these means are the use of devil-scarers and the employment of powerful spirits. A coin of Mohammad Shah's reign is washed in water, and the liquid is given to overcome the spirits that make delivery painful and long. Many objects are thus treated to secure powerful remedies through the principle of sympathetic magic. Mention has been made of incense, especially that which smells most vile. Sometimes obscene and filthy language, or food that would out-caste a respectable man, is used to drive away a demon that may have been a man of good caste. So soups are made of toads and fecal matter. Sometimes the dried tongue of a pig is put into a bag and hung from the neck. The aid of Kali is invoked through the use of an iron bracelet secured from a priest at one of her temples. Ashes from the cremation-grounds are used to drive out disease-demons.

Exorcism occupies an important place in the cure of disease. For dog-bite or snake-bite, and for the stings of various insects, nim branches are passed over the body while charms are pronounced. In this way the spirit is compelled to leave the body either through the feet or the head. Sometimes an old shoe is used to "rub down" the poison. Again, nim or mango or *madār* branches are thrown over the patient, and he is sprinkled with water from a blacksmith's shop. Red-hot iron should have been plunged into the water frequently.

The practice of disease transference is common. A wizard relates how he found in his village a yard of new white cloth, in which were tied, in separate places, seven sorts of grain and five pice. He took the cloth home and untied all the knots. He had a vest made of the cloth and spent the pice for liquor. He said that the cloth had been left by someone, at the suggestion of a wizard, because there was sickness in the house of the one who had furnished the cloth. Sometimes seven kinds of grain, seven kinds of sweets, seven pieces of haldi, and seven plantains are waved over a sick person, placed on his head, and then carried and left at a bathing-place or at cross-roads. A sick child is sometimes hung in the "cradle" of a well on a Sunday or a Tuesday; or, when the fields are being watered, a child is taken out through a breach in a trench five times in the early morning before the crows begin to caw. Diseases are transferred to a cock, a pig or a buffalo. The animal is then released in the name of some Mātā. For example, a buffalo is painted red (by a wise man) and driven through the village, with noise and the beating of drums, and then out into the jungle. If it should return the disease would break out afresh. The buffalo is purchased by a subscription, in which all freely share. Likewise goats are purchased with freewill offerings and driven out through the village. Sometimes small carts are tied to their necks for the disease-godling to ride in. At other times a male goat of one colour and without blemish is taken to the bedside of a man stricken with cholera, and the patient places his hand upon the animal. The goat is then led about the village and then to a vacant place outside the town, where it is prepared, by washing, for sacrifice. A square is marked off and plastered with cow-dung. In the square a small hole is dug to receive the blood, and a small fire of cow-dung is also kept burning. One of the elders of the community then severs the head of the animal from its body with one blow from a chopper. If the sacrificer objects to taking life, the goat is marked on the ear and turned loose. Sometimes the sacrificial animal is slaughtered in the centre of the village and a feast is held. Occasionally, the animal is tied up to keep the disease from spreading. At other times a small cart, for the Mātā, is carried through the village by the devil-priest. In this cart will be found a bit of a waterpot with black and red marks upon it, a mirror, a comb, earrings and bracelets.

In times of epidemic, such as those of plague and cholera, these disease-demons are driven from village to village with noise and incantations. At other times the devil is ensnared, by magic, into an earthen pot, the lid is put on, and it is carried off to another place. This is done, with the help of Marī, with incantations and the use of the shinbone of an ass.

It is often necessary to protect a village from such disease-demons as might be driven into it. Sometimes a magic circle is drawn, by means of milk and spirits, around the village. During epidemics people put five lights on the village boundary to bar the approach of foreign spirits, and, with much shouting and the beating of drums, drive disease-demons on to the next settlement.

Disease transference often has for its purpose the injury or destruction of others, as well as release for an epidemic. This has been illustrated in the methods of expelling plague and cholera. In cases of smallpox the clothes of a sick child are sometimes thrown behind another house, or even into another home, with the hope that the disease will go with the garments. Food is often used in the same way. Again, rags are used to transfer it to someone else.

The belief is common that Chamars use "their" demons to cause disease and for revenge.

Demons are active enemies of cattle, and cattle are protected and their diseases are treated by methods similar to those already described. For example, various forms of a ceremony known as the Gobardhan is performed at Dewali time. For increase of cattle, or for protection, a ceremony that may be classed as symbolic magic is performed. A square of cow-dung is made in which marks are drawn with barley flour. At each corner of the figure three cakes of cow-dung are piled up. Through each three cakes a broom-splint is driven; and on the top of each pile bits of cotton and cotton-seed are placed. This chauk is left in the house for three days, after which it is taken into the compound and placed in the regular runway for the cattle, that they may walk over it. The herdsman is feasted. Sometimes an offering of fire, rice and water, in honour of Mahādeo, is set out in a runway and the cattle tramp over it. To drive away cattle-disease various forms of garlands are used. For example, seven kinds of grain and two pice are put into a bag and hung over a doorway where cattle pass. Garlands of siras or mango leaves, with a mystic inscription on an earthen platter attached to the middle of the string, is hung across the village gate. At other times numbers are written on a piece of tile, with an incantation to Nona Chamari, and hung on a rope over the village cattle-path, or a wizard is employed to write a charm on a wooden label. This is hung inside a small earthen waterpot, like the clapper of a bell. This is attached to a string and hung over the village gate. It will ring when the wind blows. Or, a faqir reads a passage from some sacred book over a long string to which a red rag and potsherds with charms written upon them are attached. This is hung across the village gate.

A more elaborate device of the same kind is made as follows: To drive off cattle-disease two and a half pounds of urd and seven *chappans* (earthen jar-covers) are bound in a rope as a har, or garland. Between the chappans the urd is hung in bags. On the covers (chappans) red, yellow and black marks are made. This har is hung over the roadway by which the cattle go out to pasture. Occasionally this is hung up outside of the village, where all the animals can go under it. Cattle are branded on the legs with a circle with a cross in it, with Solomon's seal, or with Śiva's trident, or with a similar device, especially as a cure for lameness. To cure worms in cattle tie an animal in a marshy place. Tiger's flesh is burned in the cattle-stall and the smoke is "given" to the cattle, and sometimes the smoke of nim-leaves is used. Again, haldi, cloves, sugar, flowers, charcoal and dub-grass are ground together and over this limejuice is poured. A sweeper is then called. He takes his drum and leads a procession around the village, beating his drum as he goes. A man follows him scattering this preparation made of seven things as he goes, and then a young pig is sacrificed to the village godling. The disease is thus driven out. *Rorā*, a contagious disease of cattle, is transferred to the east on a Saturday or a Sunday night. No fieldwork is done, no grain is cut, no food is cooked, nor is a fire lighted during the day. The transference of rora (*rorā ḍālnā*) is carried out as follows: In procession a buffalo's skull, a small lamb, vessels of butter and milk, fire in a pan, wisps of grass, and siras sticks are carried to the boundary of the village, and then these things are thrown over the boundary. A gun is fired three times to frighten away the disease. Again, about midnight, two men carry the lower half of an earthen pot with a fire inside and a cloth beneath. They are accompanied by fifty men with long clubs (*lāṭhī*), who beat the ground and anything that they chance to meet and thus drive the disease out of each house.

The foregoing pages show how fully the Chamar enters into the practice of magic and primitive medicine, and how large a place there is for the witch and wizard, the medicine man, and the devil-priest. The legend of Nona, the Chamari, is a good illustration of many points in this belief in witchcraft. She is a deified witch much dreaded in the eastern part of the Provinces. Her name is invoked in times of trouble, in sickness, and for the cure of snake-bite. According to the legend Dhanwantri, the physician of the gods, was bitten by Takshaka, the king of snakes, and, knowing that death approached, he ordered his sons to cook and eat his body after death, so that they might thereby inherit their father's medical skill. The sons were about the eat the body, having cooked it in a cauldron, when Takshaka appeared in the form of a Brahman and warned them against this act of cannibalism. They therefore allowed the cauldron to float down the Ganges. As it floated, Nona, the Chamari, who was washing on the bank of the river, not knowing that the vessel contained human flesh took it out and ate its contents. She immediately obtained power to cure diseases, especially snake-bite. One day all the women were transplanting rice, and it was found that Nona could do as much work as all her companions put together. So they watched her. When she thought she was alone she stripped off all her clothes, muttered some spells, and threw the plants into the air. They all settled down in their proper places. Finding that she was observed she tried to escape, and as she ran the earth opened and all the water of the ricefields followed her. Thus was formed the channel of the Noni river in the Unao District.[1]

Like most people in the villages, Chamars are ever on the lookout for signs of the black art. Nearly everybody is fearful lest someone carry out sinister plots against him,

[1] Crooke, *Tribes and Castes of the North-Western Provinces and Oudh,* Vol. II., pp. 170, 171. This form of story is common. Instances of nakedness in connection with magic are numerous.

and much superstitious fear centres around the wise man and the wise woman. Belief in the magic power of human fat, or the essence of the body, Mumiai,[1] and that certain faqirs distil a medicine from the bodies of fat, black boys, is widespread. This is but a phase of the belief that results can be accomplished by mysterious powers, and that works of this nature are being done all the time. Behind this point of view is the belief not only that unseen powers are always anxious to do evil, but also that these same powers may be under the influence or power of some one skilled in the art of witchcraft.

The belief is common that certain persons are able by a look or by incantations to tear out the liver or the heart of anyone against whom they may choose to exert this influence; that they can extract substances from sick persons' bodies, and that they can throw anyone into a decline. Some witches are able to produce abortion by the use of magic squares, and others kill children by means of occult powers. Some of these uncanny women can assume the forms of beautiful young women, or of hideous *Kālīrātrī* (Black Night), or of other terrible beings, and are able to transform themselves into tigers and other wild animals. It is wise to knock out the front teeth of one suspected of being a witch, lest, in the form of a wild beast, she tear one to pieces. In the descriptions of private and of public magic and of folk medicine are found additional testimony to the belief in the exercise of occult powers.

The names denoting sorcerer, wizard, magician, exorcist, soothsayer, witch-doctor, medicine-man, and the like, are many. All of them apply to the persons occupied in various phases of the profession. The names are: *Ojhā* (teacher), *sayānā* (cunning, shrewdness, cleverness), *gyānī* (the wise one), *neotiyā*, *guṇī* (skilful, dexterous), *baigā* (an aboriginal devil-priest), *aghorā* (one who feeds upon dead bodies and other disgusting food), and *bhagat* (a devoted man). Old women who are decrepit, or of evil temper, are often

[1] Crooke, *An Introduction to the Popular Religion and Folklore of Northern India*, pp. 299, 300.

considered as witches, and some say that the power of witchcraft is possessed by persons afflicted by ugliness and bad disposition, fits and the like. On the other hand, it is said that a witch must be of pleasing disposition.

Chamars employ wizards belonging to other castes besides their own, and their own caste-wizards serve other castes, even the twice-born. Of all the "priests" or "religious" leaders of the Chamars, this class of persons is in many respects the most influential and the most dreaded. Women are more feared than men. The ignorant are kept in superstitious bondage by those persons who deal in matters more or less uncanny; and their arts and beliefs are spoken of in whispers even by the wise man himself. They live upon a spurious system of natural law, and their art is a secret service for anti-social, nefarious ends. The black art is not one which anyone can practice. It appears when the method and ends are regarded with disapproval, because injury is intended, and the purport of the act, as anti-social in spirit, is one that no one ought to perform. The magician has power all his own. While in animism man consults spirits, in magic he controls spirits. Magic exists where the man has power over nature and where he is independent of the supreme will. The wizard does everything without the aid of the gods.[1]

The black art is therefore anti-religious. The magician uses methods which include the principles of sympathy, similarity, and mimicry; and, conciously or unconsciously, use is made of suggestion, fear, fascination, telepathy, and various other psychical processes Nefarious magic is practised at night. An illustration of some of the principles involved in the practice of the art is given in the following account by a Chamar exorcist (*jhāṛ phūṁk karnewālā*): If, when a man falls ill, his condition is attributed to the activity of an evil spirit, a neotiya is called. He prepares a space, with a plaster made of cow-dung, and places the sick man upon it. A one-wicked lamp is lighted and incense is burned, and a cocoanut and

[1] Lyall, *Asiatic Studies*, First Series, p. 104.

a lime are placed in the square. Country liquor is furnished. Of this the neotiya pours out some as an oblation and then drinks of what is left. A *damru* (a drum) is then beaten in a peculiar way, for a time in slow pulsations and then very rapidly and then with varying rhythm, until the sick man is "possessed" of the spirit causing the disease; that is, until the patient begins to wag his head. This spirit is then coerced by means of a mantra, which brings the wizard's demons into action. The neotiya now asks, "Why do you give trouble?" The spirit, speaking through the sick man, states the reason, which, quite likely, may be that the spirit's grave has been befouled by the sick man. The spirit is then commanded to pronounce the words, "Rām, Rām," and to make the following promise:[1]

> "On the name of Śiva at the time established,
> I call Earth to witness,
> In whatever way he has dishonoured thee
> Let the dispute (grudge) be dropped (be forgotten),
> As when the lower half of the waterpot is broken (the pot is done for):
> Even if he (the sick man) walks backward and forward on thy breast,
> Give expression to thy displeasure not even by an "Ah!",
> Now I call for release (for the sick man),
> In the name of Prince Lachhman, of Mahādeo, and his consort, and of Hanumān, the powerful."

To escape torment at the hands of the neotiya's demons the evil spirit makes this promise and leaves the patient. The offerings which were made are considered as given to the wizard's Bir, or powerful demon, who compelled the disease-spirit to come out of the sick man. The wizard now goes home, taking with him his fee of five rupees. The cocoanut and the

[1] *Śiva bācha samo laganya*
Dhartī sākh gayā wār kī pār
Tujh ko jhagrā chhūṭ, kapṛā phūṭ
Terī chhātī chaṛhe āwe jāe
Āh to na karīhe
Ab duhāī hai
Lachhman kuṁwār, Mahādeo
Gaurā Pārbatī, Mahābīr Hanumān jī kī.

lime are divided and distributed among the people at the sick man's house. The neotiya spends the fee in the worship of some of his other godlings at his own home. To these he will offer ghi (in fire), cocoanuts, country liquor, and a goat. The head of the goat is cut off at a single stroke and placed immediately before the *deohār*. If the head "speak" he knows the offerings have been accepted by the godlings. If not, this offering is made over to some others.

For prophecy various methods to produce frenzy or trance are used. For example, while the bhagat plays the cymbals an accomplice wags his head until, finally, spirit-possession is obtained. Then the wizard tells by what spirit the person is possessed, and what the spirit wishes by way of propitiation. Frequently the sick person dances with the bhagat, and then, in a trance, tells the wizard about the spirit-possession. The demon usually expresses himself as desiring the sacrifice of a chicken, a goat, or a pig, or the offerings of sweetmeats, ornaments, clothing, or money. Of course, the bhagat receives these articles. Other means of naming spirits are used. For example, the sayana receives tobacco from the person who calls him in, which, while music is being played, he waves over the body of the patient. He then smokes, and begins to dance, and often to beat himself with a whip of cords. While in the state of frenzy which follows, he names the cause of the disease and the remedy. Or, he calls on the names of various diseases and the disease-demons as he smokes. Sooner or later he is obliged to cough, and the name that is on his lips as he begins to cough is that of the demon causing the trouble. Sometimes the wizard waves grain over the patient's body on a Saturday or a Sunday. He then counts the grains one by one and places them in heaps, and names a godling for each heap. The demon into whose heap the last grain falls is the one to be propitiated.

Simple magical practices, but with the use of mantras (charms) are common. These are used to exercise power over the one towards whom the magic (*jādū*, *ṭonā*) is directed. For this purpose, water in which the rope used to tie the front feet of an ass, or water in which a woman's napkin has been washed, is given to a person to drink. Similar secret methods used as love-charms, and for revenge, have been mentioned above.[1] It is said that owl's flesh given to either sex (secretly, of course) will make the recipient fall violently in love with the giver, and that love, so induced, is not likely to abate. So the bhagat furnishes the love-charm. Likewise owl's heart is obtained from the wizard, after proper spells are recited, in order to secure power over another. Again, an owl is killed on a Monday, and its eyes are burned. When the ashes of the right eye are thrown on a woman's garments she begins to love the one who paid for the magic. Should the man become tired of the woman, he can break the spell by using the ashes of the other eye.

The wizard rarely attempts the impossible, and invariably provides for failure ; but the superstitious attitude of mind, so pronounced among the masses, greatly enhances the influence which he is able to exert, and his directions are usually carried out minutely. Many acts, described under "Birth Customs" and under the various topics in this chapter, which seem absurd or awful, are performed because the wizard says so.

The power of the bhagat rests in the control of spirits and godlings, by means of his peculiar bhuts and divinities. These spirits, or demons, are brought under his control in one of two ways: either the spirits are given into his power by his teachers or his parents, or he secures control of spirits by means of well-known devices. Magical powers are obtained by pouring water on a babul tree for three days. By this means the person gets control of the spirits inhabiting the tree. It is said that if a person puts an owl into a room, goes in naked and feeds the bird with meat all night, he will obtain superhuman power. Such powers are obtained also by eating filth, by eating human flesh,[2] and by repeating charms backwards. Asceticism leads to

[1] Page 168.
[2] See the legend of Nona Chamari above.

such powers, which include knowledge of the past and of the future, the ability to read men's thoughts, and the power to fly in the air or to float on the water. Yogism is a system of strange, extraordinary and mysterious knowledge giving his possessor very extensive powers over men and over natural phenomena.

It is to obtain control of spirits that these persons visit the burial-grounds or burning-ghats, and make uses of ashes from the funeral pyre. Especially is this the case when a wizard, or a witch, or a woman dying in childbirth, has been burned. A man takes the flesh and liver of a black crow and cooks them separately, and likewise the flesh and liver of a pig and a goat. He then makes a karhi, places it upon a plate made of imli (tamarind) leaves, with sweets, sherbet, eggs and plantains. These are taken to the cremation-grounds, where with mantras and a fire-sacrifice they are set out for some ghost to eat. The spirit, bhut or pret, which accepts the food, thereby falls into the wizard's hands and must thereafter obey his commands.

Another illustration which has some points resembling those in the case just related, shows how a novice seeks to get control of spirits. The pupil learns the mantras from his master and follows out all instructions. He may be sent to a spot on the bank of a river where a body has been burned recently. There he washes, lights a fire, repeats spells, casts ghi, spirits and sweets into the fire, passes his hands through the fire, and touches his forehead. This ceremony he may repeat on one or two successive Sundays. His purpose is to get the spirit of the dead under his control, for then he can compel the spirit to carry out his good and evil commands.

Another method is to secure the body of a child that has been cast into the Ganges, bathe it again, dress it in new clothes, sprinkle the body with scents, put *surmā* into its eyes, light a fire and sing. His purpose is to cause the child to revive and dance. If the child come back to life the novice has its spirit in his power. If the pupil does not succeed under his master's directions, the magic is not perfect, and the process will be repeated.

Some wizards have many spirits under control, and some have also great demons (Birs) or godlings, who have a great following of lesser spirits. For the most part these demons are not widely known.

The specialists in magic transmit their lore to their pupils orally, and charms or spells are given which enable the pupil to control the spirits belonging to his master. In some cases the profession is hereditary. The pupil is most carefully trained, and in some instances he goes though a very vigorous course of discipline.

The following ceremonies were performed by a Nalchhina Chamar sayana in passing his powers on to his son: On the bank of a stream a goat, together with a lime, a cocoanut, spirits and flowers, was offered. (These things are greatly liked by the giant-demon, the Bir of Lohra, the strong one, the hero of Lohra, who is this wizard's very powerful servant.) This Bir has a host of lesser demons under his control. After the sacrifice the father led the son into the stream, and then the following protective spell addressed to the Bir was repeated[1]: "O Lord, whatever magic this one directs, that do thou bring to pass! If he makes mistakes forgive him; do not torment him." After this the mantras (charms) were taught. The mantra for Lohra kā Bīr is as follows:

"Let me control the males (men) and the powerful (the braves) and those among the jogis who are powerful.

"Let me control the five strong ones of Bengal; and the evil spirits and demons;

"Let me control the five strong ones of Prithī Siṁh (Lion of the Earth), the head-covering (and women (?) *i.e.*, who have their heads covered).

"Let me control all who live where her mother lives (persons and demons), and likewise those who live at her husband's home; and the well, and those who assemble.

"Let me control the coming and going (*i.e.*, those who pass by; or, May I have control of the streets and all who pass in them).

"Let me control the magic of the Dom and the Chamār.

"Let me control the magic of the Khatīk and the Kumhār.

[1] *Mahārāj jo kuchh jhār phūṁk kare is kā kahā karnā*
Aur agar kuchh is se bigar jāe, is ko mu'āf karnā, is ko satānā mat.

"I claim the delivering power of the powerful Hanumān, of Prince Lachhman, of Rāma and Sītā."[1]

In these lines the sayana calls upon his Bir to inhibit certain magical powers, and upon other spirits to help him in his designs. He then taught his son how to proceed in the control of spirits, and how to use mantras in his profession (*mantra phūṁknā*).

The following account shows how a witch makes human sacrifice to her demons, in order to strengthen and maintain her power over them; for the beings through whom she practises her art require human sacrifices, or at least human blood. Should she be unable to carry out some such procedure as this now to be described, she must draw blood from some person's veins, or from her own, and offer it. In the case under consideration, the demon sucks the victim's blood through the lips of the witch. When a woman who is possessed of the spirit of witchcraft has a craving to practise the art, she casts the spell of the evil eye upon a child, and utters a charm so powerful that within a certain time, which she determines—two and a half hours, or seven or fourteen days—the child dies. On the night after the body has been buried in the jungle by its parents, the witch, taking a knife, sarsoṁ oil, and a small one-wicked lamp, goes to the place of burial. Then she strips off all her clothes, and there, with her own excreta she plasters a small piece of ground. After making little balls of excreta and lighting the lamp she opens the grave, lifts out the body of the child and anoints it with oil. When the child revives, she feeds it, loves it, and plays with it until it laughs. She then places the child on the plastered spot.

[1] "*Nar bāṁdhuṁ, sūr bāṁdhuṁ, jogī bīr bāṁdhuṁ,*
Pāñch bīr Bangāle ke bāṁdhuṁ, bhūt paret bāṁdhuṁ;
Pāñch bīr Prithī Siṁh ke bāṁdhuṁ, sir kī jhagulī bāṁdhuṁ,
Maike bāṁdhuṁ, sasūre bāṁdhuṁ; kuāṁ panghat ko bāṁdhuṁ,
Awājāhī bāṁdhuṁ, Domār Chamār kī vidiyā bāṁdhuṁ,
Aur Khatīk Kumhār kī bāṁdhuṁ, Mahābīr Hanumān jī kī duhāī;
Lachhman Kuṁwār Mahābīr Hanumān jī Bhagwān (Rām) ki duhāī;
Jānkī (Sītā) jī kī duhāī."

and begins to dance. After that she cuts out its heart, sucks a little of the blood, and then replaces the heart and reburies the body. The wizard who told this explained that, if the parents suspected that the child had died of fascination, they would arrange for two men to watch in a tree near the grave that night. If one of the men should rush out and seize the witch by the tuft of hair on the crown of her head while she was dancing and cut it off, and if the other man should put out the light at the same time and snatch up the child, the witch's magic powers would be destroyed and the child would be saved alive. He cited a case in point where the witch's head was shaved and she was turned out of the village.

Magic works both ways, and bhagats are often pitted against each other. There are also devices to neutralize the effects of spells. One way is to kill a black monkey on a Thursday, drink a little of the blood, and then take the skin and wear it as a cap. If, while a wizard is curing a bewitched child, he shaves his upper leg, the other magician (witch or wizard) who cast the spell over the child will have his head shaved, and through that very act the power of witchcraft will depart from him.

The area of a bhagat's influence varies but is not very wide. His powers are greatest on the fourteenth, fifteenth and twenty-ninth of each month, at the Holi time, during the Nauratri of the Durga Puja, and during the Dewali festival. On these nights wizards and witches are supposed to be abroad. They cast off their clothes and ride tigers and other wild animals, and alligators convey them over streams.

The discovery of witches proceeds from suspicion. For example, it is noticed that, after a woman's visit to a house, a child falls ill. The same thing happens in another house. The woman is straightway under suspicion. Evidence as slight as this is frequently accepted. The case is related of a noted mahant who had a small boil on his leg. One day, as he was riding on his elephant, a woman cast an evil glance upon the leg, and very bad blood-poisoning ensued. The woman was suspected of the act, so her husband's brother was called. He spoke to her about the matter, and in reply she said, "I wish that he be well again." The swelling subsided. The narrator (a Chamar) said that this occurred about fifteen years ago. He also related that, owing to this same woman's art, a cow in her own house refused to give milk. A wizard, with powers superior to hers, who was called in, destroyed her spells, and the cow began to give milk again. But shortly afterwards the cow died. Then the woman disappeared—that is, she was murdered. The wives of four brothers in the family died within the year. This also was her work. When it is thought that there is a witch in a village, and she cannot be found out, a lamp is lighted and the names of old women are called out. The flickering of the flame as the names are being called indicates the guilty person. Other tests are used to remove doubt as to the guilt or innocence of accused persons. For example, if a wizard or a witch is struck by a branch of the castor-oil plant he will cry out. Pain is a sign of guilt. Chamars are exceedingly afraid even of a slight blow from a castor-oil switch. Another test is as follows: Two pipal-leaves, one to represent the accused and one the accuser, are chosen. These are allowed to fall upon the accused's head. If "his" fall uppermost the indications are suspicious, but if the other fall uppermost he is probably innocent. If the test leaves him under suspicion, the next day the accused is sewn up in a sack in the presence of the headman of the village, carried waist-deep into water, and let down. If he gets up in his struggles he is guilty, for a magician cannot sink in water. Another test is for a wizard who is trying to discover the sorcerer to shave the hair on his leg. The hair on the head of the witch is shaved off at the same time, and she is discovered. Sometimes witches are caught in the act. There are many instances similar to that of Nona Chamari, where the woman is discovered naked and working magic. It is said that occasionally women are found performing the following magic on the lamp-lighting day of Dewali: The witch takes four pestles (musal) into a room, and places one in each corner. She then strips off her clothes, pronounces a spell over pulse (urd) and throws it at the pestles. These leave their places, rush together with a clash, and fall down. She then puts them away. In this way she tests her magic powers. Witches are sometimes identified by the peculiar stare which is their characteristic. Some witches are accompanied by cats.

A person discovered in the practice of witchcraft or suspected of the offence is very cruelly and roughly handled. The witch is often beaten with shoes and clubs. Sometimes she is put upon an ass, set with her face towards its tail, and ridden out of the village. At other times she is compelled to work a cure where she had caused an illness. Her teeth are knocked out with a stone, or filth is thrown upon her, or she has to drink water from a tanner's vat, or her head is shaved, or her face is blackened, or mutilation is resorted to. Efforts are often made to murder her. In former times many persons were put to death every year for practising the black art, and this occurs occasionally even yet. The following is a case in point: "An extraordinary story of the murder of a woman believed to be a witch was told before Mr.—— to-day. Four coolies working at —— have been arrested in this connection. Deceased was a labourer, and recently some deaths occurred in the coolie lines. One of the accused is alleged to have said deceased was a witch, who, by her black art, had killed the coolies, and that unless the witch was got rid of more deaths would occur. About 6 p.m. on Wednesday the accused caught deceased by the hair, and, assisted by a large number of coolies, dragged her before Mr.——, demanding her ejectment from the coolie lines, on the ground that she was a witch exercising an evil influence among the coolies, with the intention of killing them. The result of the interview with Mr.—— has not been disclosed; but the coolies, being in an excited and frenzied state, dragged the woman back, brutally assaulting her on the way. Nearing the house where the deceased lived, accused, it is alleged, dashed her to the ground, killing her on the spot."[1] Similar cases are frequently brought before the courts.

[1] News item, *Pioneer*, August 25, 1910.

If, because of her particular behaviour, a visitor is suspected of practising magic, the host (or hostess) will spit upon the place where she sat, as soon as the visitor has gone. She will also stamp on the spot, and say: "If there had been a stalk of bajra, I would have chopped it up fine."[1] This should be repeated within the visitor's hearing. In case the witch is one who cannot cure, although she can cause, disease, or in case the witch is not found, a more powerful wizard or witch is called in to remove the effects of the magic. If a witch, by casting an evil glance upon the mother, causes an infant to fall ill from feeding, a wizard is called in to remove the spell. Fire is put into an earthen pot, and the mother lets a few drops of her milk fall into it. Barley husks, mustard and red peppers are placed in the pot, it is waved about the child's head, and left in the road. Cures are wrought by the punishment of witches.

The wizard, besides practising the black art, manufactures amulets and charms, and offers sacrifices to godlings and demons in times of epidemics. It is said that when such persons become Christians they lose this magical power. Neither can they accomplish their purposes when Christians are present.

The accounts given above show how thoroughly the ideas of witchcraft are planted in the mind of the Chamar. The conversations with these professionals reveal also the utter depravity of their minds. Their thoughts are full of lust and uncleanness. It is hardly necessary to remark that this kind of "religious" leader is of a low mental type, that he is bestial in his habits, and that he is given to flesh-eating and to drunkenness.

[1] *"Agar bājre kī narāī hotī, to maiṁ katar ḍāltā."*

CHAPTER VIII

HIGHER RELIGION

Some nature-gods have their places in the Chamar's religious world, but their position is not what it was in former times. Sūriyā, Sūraj Deotā (the Sun), for example, is now nothing but a godling, or perhaps a deified hero. While the Chamar is not admitted to the shrines of this godling, still, every morning as he leaves his house he bows his head, joins his hands and calls upon the Sun as Sūraj Nārāyan. There are phases of sun-worship in the domestic ritual (*e.g.*, in the phera) and in the course the cattle take in treading out the grain. Those who have no children fast and worship Sūraj Deotā, in the hope of obtaining offspring. The *swastika* and various figures in which the circle appears are symbols of the sun.

Fire also is worshipped, and is used in many parts of the domestic ritual and in sacrifice.

The stars are the spirits of people, and every person has a star. When one dies his star falls.

The new moon is addressed with the words, "Rām, Rām." On the birthday of Krishna people fast, and no one begins to eat until about midnight when the moon rises. Women worship the moon, that their children may escape disease. They take water in a lota and pour out an oblation after doing reverence to the moon. They also fast the whole day of the new moon. The moon is called māmū (uncle) by the children in Oudh. The spots on the moon represent an old woman sitting under a banyan tree, running a spinning-wheel. The halo around the moon is a sign of drought or of famine.

Eclipses are times of great anxiety, because the moon is then in great trouble. It is a time of ceremonial pollution, when bathing is necessary; and any cooked food that might have been in hand when the eclipse began must be thrown out.

An eclipse is a time when demons have the upper hand and are about in force. So buffaloes and cows with calf are marked with cow-dung and their horns are smeared with red-lead. Pregnant women are protected with marks of cow-dung (on the abdomen) and are not allowed to go to sleep. The eclipse is explained as follows: The sun and the moon were brothers. A hungry worshipper came to them, saying, "I am poor and hungry. Give me something to eat." The brothers went to a sweeper-woman, and said, "Give this man grain." She had a bin in which were all kinds of grain. She agreed to give grain to the beggar for a year. She was directed by the brothers to take the grain out of the bin from below, and they agreed to fill it by putting grain in from the top. During the year the sun and the moon were unable to fill the bin, and when the year was up, the woman said, "Now pay me, for the bin is not full." They were unable to pay her and hid themselves. Now, when eclipses occur, the worshippers of the sun and moon collect various kinds of grain, mix them and distribute them to beggars, and thus deliver the sun and the moon from shame.

Indra (Rāja Indar) is another who has fallen from his high estate. He is still worshipped in connection with the rains, but as a mere godling.

Rivers receive special consideration as great satisfiers of life. Among the sacred rivers are the Ganges, the Jumnā and the Sarjū, and their branches, the Narbada and the Sōn. Floods are caused by demons. People refrain from saving drowning folks lest they offend some deity who is thus claiming his desire. Khwājah Khizr (Rāja Kidār), the godling of the water, is worshipped by lighting lamps and feeding Brahmans, and by setting afloat on the village pond little rafts of sacred grass with lighted lamps on them. His vehicle is the fish. Water-holes are dwelling-places of demons. Wells are an object of worship, especially by women, with offerings of sweets on a tray, and by singing, and by the beating of drums. Offerings are placed on the well-curb, worshipped, and then eaten. Some, however, throw the offerings into the well. In cases of sickness none of the offering is eaten, but it is left to be taken away by some person or animal, and the consumer of the offering is expected to carry off the disease. Sometimes, at Dewali, the water of seven wells is drawn and barren women bathe in it. Chamars accept also sacred lakes and tanks as objects of worship.

The range of the Chamar's superstitions and beliefs begins with the primitive notions of animism, and reaches up through the worship of nature-deities. However, the deified saints, the tutelary, the village, and the nature-deities are not real gods, but at best mere officials of the gods, possessing but varying degrees of power. And these lesser beings are all that he knows. The great gods of the Hindu pantheon are scarcely known to the Chamar, although his beliefs are of the polytheistic type. Still, he has a vague belief of a better sort. The Superintendent of the Census of 1901, in his general report for the United Provinces,[1] said: "The general result of my inquiries is that the great majority of Hindus have a firm belief in one supreme God, called Bhagwān, Parameshwar, Īshvar or Nārāin, and that this is distinctly characteristic of the Hindus as a whole." This observation applies to the Chamar. Or, as Sir Alfred Lyall has put it,[2] the devout man trusts that there is something better beyond and above the gods. And the Chamar worships, even though it be in a hazy fashion, this Supreme Being. It would be interesting to discover whether the notions of a Supreme Being are of the Indian theistic type. This would be a reasonable conclusion, for the great movements set on foot by Rāmānanda have influenced nearly all sections of the Chamar caste, and many Chamars now call themselves after some of these reformers.

The Chamar accepts the doctrines of transmigration and of karma, and this belief explains many of the death

[1] Pp. 73, 74.
[2] *Asiatic Studies* (First Series), p. 67.

customs and some of the birth customs; and, for the most part, these ideas exercise a dumb, depressing, fatalistic influence upon them. There are, however, certain sects of the Chamars which teach that guru-worship will issue in a permanent release from the round of births.

For the most part, Chamars are denied admission to Hindu temples. Their offerings are, however, accepted; and they may stand in front of the entrance and look in. Brahmans will accept food and cash from, although they will not touch, Chamars. On the other hand, Chamars are allowed to make offerings at temples to Devi, to Bhairoṁ and to the Mātās, at some temples to Sitala, and at unenclosed temples to Śiva. In some places they have their own temples.

There are many shrines in which the Chamar has great faith, and from which miracles of healing have been reported. Such shrines are places of pilgrimage. Some of the shrines belong to local Chamar groups. Occasionally an iron chain, about three and a half feet long and weighing seven pounds, is suspended from the roof of such temples or shrines, and with this the local devil-priest may beat himself into a frenzy. There is a good deal of worship of godlings and of spirits at local village shrines and at the place where the village boundary godlings are kept. This latter place, which is generally under a tree, usually a nim tree, is made up of a heap of stones or bricks, sometimes placed upon a rude platform. Some shrines are the property of a group of villages. The images, which are mere stones, are smeared with vermilion and ghi. Sometimes clay images of horses and elephants, the vehicles of the godlings, and peculiar bowls on three legs, and beehive-shaped vessels (kalsa) are found. A small cot in the nim tree commemorates the recovery of some one from smallpox. Sometimes a devil-priest is in charge of a local shrine. The offerings before the godlings consist of lamps, cakes, milk, goats, pigs, fowls and, occasionally, a buffalo. Worship at these shrines is intermittent, and they are neglected until some pestilence or calamity falls upon the people.

Chamars employ Brahmans as astrologers. Besides this, in the west, nearly all Chamars employ Chamarwā Brahmans as priests. In the east, the well-to-do engage degraded Sarwariyā or Kanaujiyā Brahmans. Gurrā Brahmans, who wear the sacred thread but who are looked upon by other Brahmans as polluted, receive offerings from Chamars but do not eat with their clients nor enter their houses. Chamarwā Brahmans serve certain sub-castes of Chamars and sometimes preside at weddings. Formerly they intermarried with Chamars,[1] but they are now an endogamous group. Those of the Punjab were Chamars. A branch of the Gaur Brahmans, the Chamar Gaudas, serve the Chamars as priests. The Jatiyas of some parts of the Punjab employ high-caste Gaur Brahmans.[2]

Chamars have their *bairāgīs* and sadhus, and these mingle with other mendicants at such places as Jagannath, and often bear the brand-marks of Dwarka, Badrinath and Jagannath.

Another and important class of religious leaders are teachers, or gurus, men of various sects, who travel over the country expounding religious doctrines and initiating candidates into their special *panths* (sects), and who have a comparatively good influence upon the community. They are held in high esteem and are usually obeyed. The more respected and better instructed men amongst them, who are accepted as leaders and who are honoured by the title of guru, derive their support from offerings and fees. The following statement, abridged from Crooke's *Tribes and Castes of the North-Western Provinces and Oudh*, is a description of the travelling Kabīr Panth mahant, or guru: When a disciple is initiated by a guru of the sect, a piece of ground in the house of the candidate is plastered with cow-dung. On this spot is placed a pitcher full of water. In the mouth of the pitcher mango-twigs are fixed. On the pitcher a lamp containing ghi is lighted and an offering consisting of sandal-wood, holy rice,

[1] Rose, *A Glossary of Tribes and Castes of the Punjab and North-West Frontier Province*, Vol. II. p. 131.

[2] *Encyclopædia of Religion and Ethics*, Vol. III. p. 353.

flowers and incense is burned. A garland of flowers is placed around the neck of the pitcher, and the core of a cocoanut with some batasas is offered. Camphor is burnt. The candidate sits in the holy space before the guru, who says to him, "Repeat the name of the true being within you with breath." Then morning and evening prayers are taught. The initiate is taught also a number of hymns to be sung morning and evening. The guru visits his disciple once a year, in the cold season, and he and other mendicants of the sect are entertained by him for a couple of days. Every day the disciple washes the big-toe of the guru and drinks the water, *charanāmṛita*. The disciple then, with his hands joined, thrice makes obeisance to the guru and utters thrice the word, "Bandagī, Sāhib." As long as the guru remains in the house the disciple joins with him and his mendicants in singing songs morning and evening. When the guru is leaving, the disciple does obeisance, makes him a present of money, vessels and other useful articles, and salutes him with the words "Bandagi, Sahib." When the disciple visits his guru he is entertained by his teacher, but he leaves a present when he departs. Everything important in the life of the disciple is subjected to the control of the guru. "The ordinary mahants are not men of great learning, though they have usually committed to memory a certain number of sayings attributed to Kabīr, and possibly also some book of which they have managed to secure a copy. Want of learning is in some sort atoned for in the opinion of their followers by a detailed knowledge of the ritual to be observed in the performance of religious ceremonies. The more learned mahants have some knowledge of Tulsī Dās's Rāmāyan and of the Bhagavad Gītā."[1]

In the great sects the guru is worshipped as a god. The dust of his feet is believed to convey spirituality, and the water in which he has washed his feet is drunk by disciples as a nectar for the soul.

Besides the travelling gurus there are a number of famous poets and teachers who are reverenced by

[1] Westcott, *Kabīr and the Kabīr Panth*, p. 120.

the Chamars, and counted as gurus in a much higher sense. Among these are Vālmīki, the low-caste author of the Rāmāyana; Tulsī Dās, the author of the modern popular Rāmāyan; Sūr Dās, the blind poet of Agra, who put into poetical form the legends of Krishna; Nārad, after whom one of the Purāṇas is named, and who is connected with the legends about the birth of Krishna; Kālī Dās, the great poet of the fifth[1] century of our era; and Vyāsa, the reputed sage and author. Still others, the founders of those sectarian movements which have gained lodgment amongst the Chamars, are worshipped as gurus. In these guru-worshipping sects we find the highest types of religious life that prevail in the caste.

The movement which Rāmānuja started in South India was carried into Northern India by his great disciple Rāmānanda. He brought with him the conception of God as a person who cares for all men and who rewards their devotion. He also spread the revolt against caste-exclusion, insisting that men of low caste, and even the "untouchables," are capable of spiritual religion (*bhaktī*). He, who himself had been outcasted for the supposed violation of the strict rules of commensality, became the missionary of popular Vaishṇavism in all Northern India, preaching the worship of Vishṇu under the name of Rāma.

Perhaps the greatest of Rāmānanda's disciples was Kabīr.[2] He grew up in the home of a Mohammedan weaver (Julāha), but he came under the influence of Rāmānanda, and afterwards became his disciple. Through Kabīr Mohammedan elements were brought into the theistic movement, and by him the process of emancipation from the strictness of Hindu thought and caste were carried much further than they had been by Rāmānanda. The real importance of Kabīr rests in the enormous influence which he has exercised upon subsequent religious thinking, specially as it has affected the masses,

[1] Ryder, *Kalidasa, Translations of Shakuntala, &c.*, Introduction.

[2] See Monier-Williams, *Brahmanism and Hinduism*, pp. 158 ff; Westcott, *Kabīr and the Kabīr Panth*; Ahmad Shah, *The Bījak of Kabīr*.

because of his use of the vernacular. His attitude towards caste drew to himself a large following from the lower levels of society. He left twelve distinguished disciples, nearly all of whom were of low-caste origin, and each of whom founded an independent order. The Satnāmīs, the Dādū Panthīs, the Śiv Nārāyans, and the Malūk Dāsīs trace their origin to Kabīr, or to his teaching and influence. Nānak, Sūr Dās, and Tulsī Dās owe much to him; and a large part of the Ādi Granth is his. His influence is by no means confined to these limits, but these are the names with which this chapter is immediately concerned.

Kabīr died at Maghar, in Gorakhpur, at an advanced age. Hindus and Mohammedans claimed his body. The disputants were about to resort to blows, when an aged man appeared at whose command they lifted the sheet which covered the body. There they found nothing but flowers. These they divided. The Mohammedans took half and buried them at Maghar. The Hindus carried the remainder of the flowers to Benares, where they burned them, and then buried the ashes at the Kabīr Chaurā.

The teachings of Kabīr are found in the *Bījak*, the *Sukh Nidhān*, and the *Ādi Granth*. To-day the *Bījak* is one of the most popular literary collections in Northern India. "His best hymns are probably the loftiest works in the Hindustani language, and hundreds of his briefer utterances have laid hold of the common heart of Hindustan." In 1901 the Kabīr Panthīs numbered 850,000, of whom 500,000 were found in the Central Provinces. In 1911 there were 600,000 in the Central Provinces. This great following now stands midway between idolatry and monotheism. Kabīr Panthīs are better known in the Ganges valley and in the Central Provinces than in the Punjab. They are largely recruited from the Chamars and the Julahas. Nowadays the members of the sect are divided on caste lines, which are not broken except in the presence of the chief guru on the anniversary of the birth of Kabīr, and amongst the lower castes of the sect. "All who desire to become members of the Panth are required to renounce polytheism and to acknowledge their belief in one only God (Parameshwar). They must also promise to eat no meat and drink no wine; to bathe daily and to sing hymns to God both morning and evening; to forgive those who trespass against them up to three times; to avoid the company of all women of bad character, and all unseemly jesting in connection with many subjects; never to turn away from their houses their lawful wife; never to tell lies; never to conceal the property of another man; and never to bear false witness against a neighbour, or speak evil of another on hearsay evidence."[1]

Members who renounce the world and attach themselves permanently to the monasteries belonging to the order are called bairagis. Women as well as men may become ascetics.

The Kabīr Panthīs of Northern and Central India are divided into two branches, with headquarters at Benares, with a branch at Maghar, and at Kawardha and Damakheda in the Central Provinces. The monasteries at these places are in charge of mahants. Under these there are a number of branch establishments, also under mahants.[2]

The travelling gurus, or mahants, are recruited from various castes, and usually serve those from whose caste they have come.

Another great leader whose influence has been profoundly felt by the Chamar was Nānak,[3] the founder of the Sikh movement. Both Nānak and his successors are counted as gurus. The great guru, however, after Kabīr is Nānak. He belongs to the movement that produced Kabīr. He was a great traveller, who taught by means of hymns and aphorisms. In his earliest years he showed wonderful precocity in the acquisition of knowledge. Later he refused wealth in order to become

[1] Westcott, *Kabīr and the Kabīr Panth*, pp. 112, 113.

[2] For a detailed account of the Panth see Westcott, *Kabīr and the Kabīr Panth*, Chapters V. and VI.

[3] See Monier-Williams, *Brahmanism and Hinduism*, pp. 161 ff.; Russell, *The Tribes and Castes of the Central Provinces of India*, Vol. I. pp. 277 ff.; Macauliffe, *The Sikh Religion*, Vol. I. pp. xl. ff.

a religious mendicant. He too emphasized the teaching that men of all castes and races can know and love God. But his anti-caste principles were compromised somewhat by his admitting the lower castes on an inferior footing, and caste has found its way into the Sikh community.

Nānak too emphasized the duty of obedience to religious teachers. He revolted against ceremonial and social restrictions and priestcraft, though in no violent way. He did not go nearly as far in reform as Kabīr did. He taught that God is neither Allah nor Parmeshwar, but the God of the whole universe, of all mankind, and of all religions. Although he spoke of God as personal, he also talked of him in terms which resemble the teaching of the Vedanta, and he was much closer to Hinduism than was Kabīr. For Nānak salvation consisted in repentance and in true, righteous conduct, perfection being the end of a long process involving transmigration. He insisted upon a quiet but profound religious life, and tried to make it attractive.

Amongst the Sikhs are found *simhs*, those who are distinguished by the five "K's," and who are constituted by initiation. These are one of the later developments of the Sikh movement. The ordinary Nānak Panthīs, however, are distinguished by no peculiar customs, but they revere the *Ādi Granth* as do other Sikhs. A section of Sikh Chamars is known as the Rām Dāsīs. These are often confused with the Rāe Dāsīs.

The two large followings of the Sikhs amongst the Chamars belong to these two classes: the Kabīr Panth and the Nānak Panth. These Chamars with Sikh affinities are amongst the most enlightened groups in the caste, and with them idols and idolatry are almost unknown.

One of the most noted of the followers of Rāmānanda was the Chamar, Rāe Dās, or Ravi Dās.[1] A considerable amount of legendary matter has arisen concerning him,

[1] See Crooke, *Tribes and Castes of the North-Western Provinces and Oudh*, Vol. II. pp. 185 ff.; *Rāe Dāsī Kī Bānī*, Belvedere Press, Allahabad, 1908; *The Religious Sects of the Hindus* (C.L.S.), p. 57; Macauliffe, *The Sikh Religion*, Vol. VI. pp. 316 ff.

and in some legends an effort is made to give him a respectable ancestry. In one account he is represented as a Brahman reborn from the womb of a Chamari. The story goes that a Brahman disciple of Rāmānanda used daily to receive necessary alms from the houses of five Brahmans. This was cooked by his preceptor, and offered to the Creator before being eaten. One day, on account of rain, the Brahmachārī accepted supplies from a Baniya. When Rāmānanda had cooked the food, the Divine Light refused to accept it, because it was unclean. Inquiries revealed the fact that the Baniya had money dealings with the Chamars, and that the food was in consequence defiled. Rāmānanda in anger commanded that the disciple be reborn from the womb of a Chamari. When the infant was born, remembering his past life, he refused to suck from the breast of his mother, because she had not been initiated into Rāmānanda's sect. She was still only a Chamari. Heaven commanded Rāmānanda to initiate the whole family and then the infant consented to be fed. The child was named Rāe Dās.

At the age of eighteen this young Chamar began to worship a clay image of Rām and Jānkī. His father, displeased with him, turned him out-of-doors. Rāe Dās set up in business as a shoemaker, and worshipped as before. He made a practice of giving shoes to all wandering ascetics. One day, seeing his unusual asceticism, a wandering saint gave him a philosopher's stone. Rāe Dās paid no attention to it; for he said, "God only is important, and to use his name is the only good." But the saint touched his shoemaker's knife with the stone, and the knife turned to gold. The sadhu then left the stone in the thatch of Rāe Dās's house. Rāe Dās refused to use the stone. After thirteen months, Vishṇu, disguised as a saint, returned, and seeing the stone still in the thatch, showered gold upon Rāe Dās. Still the shoemaker did not accept the offered wealth, for he was afraid of riches. Afterwards Krishna appeared to him in a dream, and said, "Use the gold for yourself or for God." Then Rāe Dās accepted the stone, built a magnificent temple, and established regular worship. Enraged Brahmans appealed to a

Raja against him. Summoned before the king, Rāe Dās was commanded to exhibit his miraculous powers. He could perform but one miracle. At his command the *sāligrām* would leave its place and come into his hands. The Brahmans could not do likewise. A Rani thereupon became his disciple. When she returned home from her pilgrimage to Benares, she gave a feast and invited Rāe Dās; but the Brahmans, refusing to eat in the palace, took fresh grain into the garden and cooked it there. Then, while they were eating, suddenly they saw Rāe Dās sitting and eating between each two of them. They fell at his feet repentant. He then cut his skin, and showed them under it his Brahmanical thread, thus proving himself to have been a Brahman in his previous life.

In another legend it is reported that a well-to-do man of good caste went to see the famous Rāe Dās. When he reached the dwelling-place of the guru, he saw a venerable Chamar with a group of shoemakers busy making shoes. After the interview with Rāe Dās, a Chamar brought, in a large shoe, water in which the feet of Rāe Dās had been washed, and each partook of it. The visitor received the nectar, but threw it over his head. Some of the water fell on his coat and dried there. When he returned home he took great pains to purify himself from the contamination which had resulted from his intercourse with the Chamars. He gave the clothes which he had worn to a sweeper. The sweeper thereupon was transformed in a wonderful way. But the rich man became a leper. After much unsuccessful doctoring, he returned to Rāe Dās, with the hope of receiving more nectar (charanamrit). But he was disappointed. He then besought Rāe Dās to have mercy upon him, and, finally, his request was granted and he was cured of his leprosy.

Another legend, which has some distinctly Chamar characteristics about it, relates the origin of a lasting record of his unfaithfulness on one occasion. The story is that one day a cow died, and the owner came and asked Rāe Dās to remove it. He, with the help of God (Bhagwān), came and carried the carcass away. The flesh was divided, but Rāe Dās hid the heart in the ground. Afterwards when Bhagwān asked him if he had hidden any part of the carcass Rāe Dās answered, "No." Immediately there sprung up a new species of plantain whose flower took the form and colour of a heart.

Another story has it, that once, in Benares, a Brahman used to make offerings to the Ganges for a certain warrior. One day this Brahman went to Rāe Dās's shop to buy a pair of shoes. While he was there he talked with Rāe Dās about the worship of the Ganges. Rāe Dās said to him, "I will give you this pair of shoes. Will you please offer this betel-nut to the Ganges for me?" The Brahman put the nut in his pocket, and when he went to the Ganges again he made an offering for his warrior-friend, but forgot that which Rāe Dās had given him. When he was returning from the river, he thought of the betel and went back and threw it into the Ganges. But he saw that Ganga raised her hand from the river and received the offering. The Brahman perceived the meaning of this act on the part of Mother Ganges.

Followers of Rāe Dās believe that at the age of one hundred and twenty years he reached *Brampad*, the state of bliss, and then disappeared in the flesh. He took his sayings (*Bānī*) with him.

Rāe Dās, who was born at Benares late in the fifteenth century, came under the influence of the great Rāmānanda, and afterwards became the founder of a widespread religious movement. He was a monotheist, following the general lines of his master's teaching, and of even purer faith than Kabīr. The influence of his teaching has been sufficiently great to give him the place of a teacher (Brahmachārī) in the *Bhaktā Mālā* (Lives of Vishṇū Saints). Followers of Rāe Dās, amongst whom are a great many Chamars, are found all over the Provinces. Many Chamars prefer to be known as Rāe Dāsīs. Members of the sect are very numerous in the Punjab also, especially in the Gurgaon, Rohtak, and Delhi Districts, where they are all Chamars. In those areas they have increased considerably in numbers during the last twenty years. In Gujrat they are known as Ravi Dāsīs.

Rāe Dās taught that the soul differs from God only in that it is encumbered with a body. For him God was everything, and he gave himself over to passionate devotion to the Deity, believing that God is gracious to all and is accessible to persons of lowly birth. God alone can save a man from evil passions. His conceptions are based on the general principles that underlie the teachings of all the reformers.

An important unitarian sect, the Śiv Nārāyanas[1], owe much to the same sources that produced Rāe Dāsa's movement, and their opinions are somewhat similar to those of the older sect. The founder of this notable movement was Śiv Narāyan, a Rajput who was born in the eastern part of the United Provinces.

There were in the United Provinces, in 1901, 46,727 adherents of this sect. Persons of any caste may join the Śiv Nārāyanas, but Chamars, notably Jaiswars and Dusadhs, number many more than any other caste. Those who wish to become members of this religious body are brought to a *sant*, who teaches them the moral precepts of the order. Truth, abstinence from spirituous liquors, honesty, mercy and charity, even in look, are cardinal virtues of the sect. Polygamy is prohibited. Sectarian marks are not used, but conformity to the external observances of Hindus or Mohammedans, independently of religious rites, is recommended. Practice is often far below the level of their ideals, and Śiv Nārāyanas of the lower orders are occasionally addicted to drunkenness.

When a candidate wishes to affiliate himself with the order, he is first warned of the difficulties before him and is tested for a few days. If he is then approved, he is directed to bring a present, according to his means, to the *Bijak* (their sacred book). He then makes his choice of a guru, or sant, from amongst those who are present in the assembly. This sant, who sits with the scriptures opposite him, first makes in behalf of the candidate a sacrifice by burning camphor and *dasoī* (ten kinds of perfumes). Then some camphor is burnt before the scriptures, and

[1] See *Religious Sects of the Hindus* (C.L.S.), p. 147.

all present rub the smoke over their faces. The candidate then washes the big-toe of his teacher and drinks the water (charanamrit). Next, the sant whispers into his ear the formula (mantra) of initiation. This mantra is concealed carefully from outsiders. The initiate now distributes sweets to the congregation. He is then considered a sant, or initiate, and receives a small book (*parwāna*), which he is permitted to study, and which serves as a pass of admission to future meetings. If he lose his parwana he may obtain another on payment of a small fee. The parwana contains a few teachings which will be most helpful to a man in his daily life, and the little book is valuable in the hour of death, for, if a sant die away from home, this book will be found upon his person, and his own sect-fellows will perform his funeral rites.

The title "bhagat," which is taken by some sants, simply implies that they are monotheists. In some parts of India the Jaiswar groom is known as "Bhagat Sais." Amongst other duties, the sant makes arrangements for funerals, for the processions and for the carrying of the body, and sings the funeral hymns and reads the scripture on the way to the grave. Those sants who have disciples are called gurus. Those who are well-informed become sadhus, or mahants, but still continue to be householders. They become a higher order of religious leaders, who direct services in the meeting-houses, make the sacrificial offerings and distribute the *prasād* (the food of which the teacher has partaken).

Their chief monasteries are found at Chandrawar, Bhelsari, and Sasra Bohadpur, in the Ballaja District, and at Ghazipur. Their meeting-houses are known as Dhāmghar (House of Praise) and sometimes as Somghar (House of Meeting), or Girjāghar (Church). These are found in various places. In them are found usually pictures of the saints Gorakhnāth, Rāe Dās, Kabīr Dās, Sūr Dās, and others. The chief object of interest in the Dhāmghar is the scriptures, which are kept rolled up in a cloth on a table at the east. The scriptures are worshipped. They are carefully watched, and no one but members of their own

DEVIL PRIEST, AHAWAR CHAMĀR

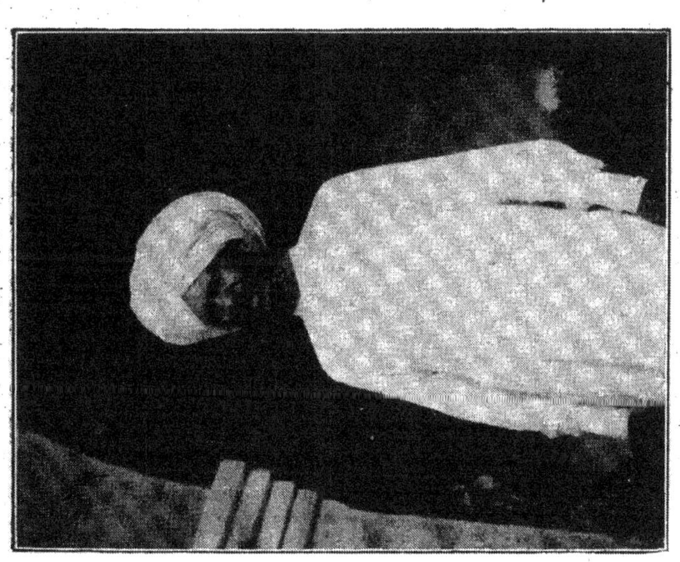

ŚIV NĀRĀYAN MAHANT, JAISWAR CHAMĀR

KURIL CHAMĀR

JAISWAR CHAMĀR

congregation is allowed to read them. Meetings are held on Friday evenings, and any educated man (mahant) among them may read and expound passages from the *Gurunyāsa*. After the mahant has finished his reading, he receives the contributions of the faithful. At these meetings there is music and singing, worship, reading from the *Granthas*, and instruction in the teachings of Śiv Nārāyan. Men and women sit apart. Members are not allowed to eat meat or drink spirits before going to the weekly service, and in the Dhāmghar they are not allowed to drink, but they may smoke *gāṁjā* (hemp), *bhaṁg* (hemp-leaves) or tobacco there. A special meeting is held on the Basant Panchami, or fifth light half of Māgh. A halwai is called in, who cooks some halwa (which is known as *mohanbhog*) in a large boiler (*karhāo*). This is first offered to Śiv Nārāyan before the scriptures of the sect. Until this is done no Chamar is allowed to touch it. The explanation of this is that Śiv Nārāyan was a Kshatri, and it would be defilement to him if any Chamar touched it before dedication; and, besides this, many castes are represented at the feast.

'Śiv Nārāyans claim that their sacred scriptures have existed for more than eleven hundred years, but that they were unintelligible until they were translated by an inspired sanyasi. The present recension is the work of the Rajput Śiv Nārāyan of Ghazipur, who wrote in the first half of the eighteenth century. Their *Granthas*, or Scriptures, number sixteen, of which the most important are the *Guranyāsa* and the *Santvirāsa;* the former consists of selections from the Purāṇas, and the latter is a treatise on morals. The *Santvirāsa* is read only at funerals, where it is recited from the moment of dissolution until the burial has been completed. The *Guranyāsa* is read in their religious meetings.

The teachings of the sect are of the same type as those of the other reforming Vaishṇavite bodies; but some claim that the movement owes much to Christian influences. Some village Śiv Nārāyans assert that they worship Jesus Christ under the name *Dukhāran Guru*, or "Trouble-Chasing Guru." Members of this sect are much more friendly towards Christianity and are more easy of access than orthodox Hindus. They claim to worship one God, of whom no attributes are predicated; and they offer no worship, nor do they pay any regard whatever to any of the objects of Hindus or of Mohammendan veneration.

Another noted founder of a sect was Dādū (Dādū Dayāl Jī)[1], a cotton-cleaner of Ajmer. He was born at Ahmedabad, and lived at Sambhar and at Amber as well as at Ajmer. Dādū was rescued from a river as an infant and given to a Brahman who had begged the boon of a son from a holy man. It is said that when Dādū was eleven years old a sadhu came to him and offered to teach him, but Dādū did not recognize the man and allowed him to depart. Seven years later the holy man returned and led Dādū into the life of an ascetic. He became a man of such compassion that he was called Dādū the Merciful. Legends relate how he refused to return evil for evil. He became famous as a worker of miracles also. Though an ignorant man, he became a spiritual and social reformer. Tradition has it that he received command by revelation to become a religious leader. When he was about thirty years of age he went to Sambhar, where he lived for six years; he then moved to Amber; fourteen years later he began to travel; and after ten years he died. Dādū did not die like ordinary men, but disappeared from the world in accordance with a message that he received from Heaven, and the place of his disappearance in the Rajputana hills is still shown. His followers believe that he was absorbed into Brahma.

The sect is really an offshoot from the Sikh movement, and is sometimes said to be identical with the Nānak Panth. Dādū Panthīs believe in the unity of God and worship him under the title "True God" (*Sat Rām*). Their worship is restricted to the repetition of the name "Rām" and of the name "Dādū Rām." Still, their God is of the impersonal Vedantic type. They believe in evil spirits. The worship of idols is forbidden; but they

[1] See *Religious Sects of the Hindus* (C.L.S.), p. 53; *Dādū Dayāl Kī Bānī*, Belvedere Press, Allahabad, 1914; *Encyclopædia of Religion and Ethics*, Vol. IV. pp. 385 f.

worship the *Dādū Bānī* and prostrate themselves before Dādū's sandals and old clothes.

In earlier times they built no temples; but now there are temples in which their sacred book is worshipped. They believe that perfect devotion results in union with the Deity; but that imperfect devotion does not break the round of transmigration. Their chief place of worship is at Narana, near Jaipore. Here Dādū's bed is preserved and his books and clothes are kept. They now have a number of places (Dādūdwāra), which combine a monastery with a preaching-place, because the services in their meeting-houses are conducted by their sadhus. In the prayer-room is found a manuscript copy of the *Bānī*. In worship lamps and flowers are used. There are two main divisions of the followers of Dādū, the *sevaks* and the *sadhus*. The former are householders and men of affairs, and they are not counted as true Dādū Panthīs. They are allowed to read the *Bānī*. The sadhus are divided into four orders. These sadhus are all celibates, and they may be either men or women. The most interesting of these sadhus are the Nāgās, who serve as soldiers. The initiatory ceremony of their gurus is simple. Those of their sadhus who are able to learn are taught to read and are instructed in the tenets of the sect. They are also required to memorize the twenty-four guru mantras (which refer to the character of God), and the pañcharatī (which are used in the praise of God). They carry beads in their hands. The only peculiarity in their mode of dress is a four-cornered or round white skull-cap with a flap hanging down behind. They do not use sectarian marks, nor do they wear rosaries, but they carry beads (*sumarnī*) in their hands.

The sacred books of the sect are the *Dādū Bānī*, the *Sukhyā Granth*, and the *Janam Līlā*.

Many low-caste followers, including Chamars (some of whom are Balaīs), have been attracted to the movement through Garīb Dās, one of Dādū's disciples, but they are not admitted to the temples. Although Dādū thoroughly organized his movement, it is now on the decline. A few persons have recently withdrawn, under the name of Benāmī. They use no name in the worship of God.

Among the lesser sects which have a following amongst the Chamars may be mentioned the Malūk Dāsīs, the Lālgīrīs, the Ghīsa Panthīs, and the Rām Rāmīs. Malūk Dās,[1] who was born at Kara and who died as Puri, lived during the time of Aurangzeb. He was probably a trader. A great deal of legendary material, praising the wonderful things that he did because of his great kindness and mercy, has been preserved. When he was a child of but five years playing in the streets, he collected the thorns out of the dust so that people might not step on them; and while he was thus engaged, a great saint who happened to pass by prophesied that Malūk was destined for some great life, either that of a prince or of a saint. From his youth up he paid great attention to travelling teachers, and many stories are told concerning his care of wandering ascetics. At the age of ten or eleven years he was started in business with a wholesale dealer in blankets. He used to go into the country regularly to sell blankets, but he always gave to sadhus and to the poor what they asked. On one of his journeys, when he had made no sales and had met no beggars, he sat down under a nim tree late in the day to rest. His load was very heavy. A labourer came along and offered to carry the load for two pice. To this Malūk agreed, sent the coolie on ahead, and gave the blankets no further thought. When the porter brought the blankets Malūk's mother doubted his story, and on the pretext of giving him some food, led him into one of the rooms of the house and then locked the door. When Malūk reached home, his mother scolded him for his carelessness, and ordered him to count his blankets to make sure that none were missing. When Malūk opened the door of the room where the coolie had been shut up, he found that the man was gone. He had left behind a piece of bread, which Malūk received as prasad. He remarked to his mother that she had been very fortunate indeed to escape without a curse. He saw in that coolie a vision of God, and confessed that he had

[1] See *Religious Sects of the Hindus* (C.L.S.), p. 51; Bhattacharji, *Hindu Castes and Sects*, p. 446; Growse, *Muttra*, p. 230; *Malūk Dās Kī Bānī*, Belvedere Press, Allahabad.

not known who the man was. So he entered the room where the coolie had been confined, and directed his mother not to disturb him until called her. After three days of meditation he had a vision, came out and saluted his mother. From that time he practised meditation, and his fame began to spread in all directions. Many began to come to see him, many spiritual blessings were obtained from him, and he began to exhibit miraculous powers. There are many legends which deal with his wonderful works. He started out one time to beseech Indra to give rain during a great famine, but one of his disciples made the great god so ashamed of himself that he gave the rain before Malūk reached the fields. Another legend, growing out of this, illustrates Malūk's simple-mindedness and humility. Later, Malūk was summoned by the Emperor to Dehli. He appeared before the ruler, having made the journey by the exercise of his miraculous powers. Several dishes of khichri were prepared for Malūk, but the first turned out to be an abomination; the second one proved to be ashes, from which he raised such a dust-storm that it threatened to destroy the city, and it was only through the intercession of the emperor, and then only through Malūk's miraculous powers, that Dehli was saved. Malūk performed another miracle, in which he stood in the midst of a well without any support. The emperor was so impressed with Malūk's sainthood that he offered him gifts. His request was a simple one, which saved the officers who had been sent to bring Malūk, and one which convinced them also of his divine powers. One of the officers became a disciple of Malūk. Other miracles are recorded, such as his saving the workmen who had been buried by a falling house, and his bringing a milkwoman's son back to life. His days were filled with wonderful deeds. He died at the age of 108 at Puri. His tomb is at Kara, near Allahabad. On the day of his death he told his disciples that at noon they would hear the sound of a bell and of a horn in their hearts, and that this would indicate that he had died. He directed them not to burn his body but to consign it to the Ganges. The body floated down to Prayag-ghat

(Allahabad). There it asked a ferryman for a drink and then sank. It next appeared at Kaśi (Benares). There it asked for water and pen and ink. With these it wrote, "I have reached Kaśi." It then sank again, and reappeared at Jagannath Puri. Jagannāth Jī showed his disciples, in a vision, a car (*rathī*) on the seashore, and ordered them to bring it and place it before his image. They did as he directed, and left the car before the image and retired. The temple doors thereupon closed of themselves. Then Malūk Dās, who was in the car, requested a place to rest under the eaves of the temple and the refuse food from the temple. He (Malūk Dās) received the scum from the cooked rice and dal for his bread and parings of vegetables as a karhi. Malūk Dās's resting-place is still found at Jagannath Puri, and there "his" bread is still used and offered to pilgrims.

Six months before he died he named his nephew as his successor. Although he was a householder, he founded a monastic order. Their principal monastery is at Kara, on the Ganges. Other monasteries are situated at Benares, Allahabad, Lucknow, Ajudhya, Brinbaban, Patna, Jaipur and Puri. Still others are found in Gujrat, Multan, Nepal and Afghanistan. His followers hold no distinctive teachings, being members of one of the Sītā-Rām worshipping sects which sprang from Rāmānanda, but they take Malūk Dās as their guru. Their sectarian mark is a single red line on the forehead. Most Malūk Dāsīs are householders.

Malūk is said to have written a Hindi poem, the *Dasratha*, and a few short *Sākhīs* and *Padas*, but none of these have been published.

In the earlier part of the nineteenth century a Chamar, Lālgīr[1] by name, founded a sect known as the Lālgīr Panthīs, or Alakgīrīs. His home was in Bikaneer. According to Lālgīr's teachings men should forsake idolatry, practise charity, avoid taking life, abstain from the eating of meat, and practise asceticism. He denied the possibility of a future life, taught that heaven and hell are within,

[1] See Sherring, *Hindu Tribes and Castes*, Vol. III. p. 62.

and insisted that all ends with the dissolution of the body. He held that the ends for which a man should practise virtue are peace in life and a good name after death. The sole worship of the sect consists in calling upon the incomprehensible God, "Alakh, Alakh," and it is from this practice that the sect is sometimes named.

Ghīsa, a Jat,[1] was born at Kekra in the Meerut District about the middle of the nineteenth century, and died about twenty years ago. (Some put him a little earlier.) He began as a worshipper of Kabīr. Later he attracted to himself a considerable following, chiefly Chamars and Julahas, and formed an independent sect. He forbade animal sacrifices and idolatry. His followers are called *sādhs*, and they wear a rosary of Kāthwood beads. Once a year Ghīsa Panthīs visit their gurus, bringing gifts, and have a feast. The sayings of Ghīsa have not been reduced to writing. His teachings do not differ from those of Kabīr.

Kālū Bāba, or Kālū Kahār, or Kālū Bīr, was the founder of a sect amongst whose members are some Chamars. He discovered, by accident, that being a sadhu was more remunerative than following his usual avocation. His followers have much the same beliefs as the Sikhs; but they are at the same time worshippers of Krishna and devotees of Śiva. They reverence the Grantha. Kālū is sometimes spoken of as a low-caste godling worshipped by Chamars and others of low degree.

The Rām Rāmīs are a small group of Chamars who organized about thirty years ago. They are found chiefly on the south side of the Mahānadī, in the Central Provinces. They carry a flute, put peacock feathers around their caps, and cry out "Rām, Rām." They mean always to keep Rām in mind. Their most distinguishing characteristic is that they have the couplet "Rām, Rām" tattooed all over their bodies.

The Satnāmī movements have their rise in teachings of Kabīr. The word means the "True Name" and indicates

[1] Another report has it that he was a weaver. See *Census Report, Punjab*, 1911, p. 144.

that they worship the One Reality under this title. The first-known movement bearing this name appeared in the seventeenth century at Narnal, seventy-five miles south-west of Delhi. The sect had a reputation for esoteric doctrines, and for uncleanness in morals and in eating. They came into conflict with the Government of Aurangzeb, and were sanguinarily overwhelmed in 1673.[1]

A sect by the same name (Satnāmī) appeared in the next century, but there is no evidence to show that it was a revival of the earlier movement. Its founder was Jagjīwan Dās,[2] a Thakur, born in a village not far from Lucknow. His father was a farmer. From childhood he showed an interest in higher things and he associated much with sadhus. One day a most holy faqir, Bullā Sāhib, in company with a still more holy man, Govind Sāhib, stopped where Jagjīwan was grazing cattle. He hastened to fulfil the request for fire for their pipes (chilam) and at the same time brought milk for them to drink, although he was afraid that his father would punish him. Bullā, the saint, read his thoughts, and comforted him, saying that there would be no less milk but more at home, in spite of his having brought some for them. And, sure enough, when Jagjīwan Dās went home, he found all the pans full to overflowing. He then ran after the saints and begged to be accepted as a disciple and to be initiated. Through the compassion of Govind Sāhib he was transformed into a man of deep love and austerity. Bullā then stated that the object of their visit was to arouse Jagjīwan Dās, who, he said, was, in a previous life, an ascetic of renown. The sadhu prophesied that ere long Jagjīwan would become an expert recluse (*pūrā jog*). Jagjīwan asked for a sign to prove that this holy man spoke truth. Thereupon Bullā

[1] See J. N. Sirkar in *The Modern Review*, 1916, p. 385.

[2] *Oudh Gazetteer* (1877), Vol. I. pp. 361 ff; Russell, *The Tribes and Castes of the Central Provincies of India*, Vol. I. pp. 307 ff; *Religious Sects of the Hindus* (C.L.S.) pp. 146, 147; *Indian Antiquary*, VIII. pp. 289 ff: Prasadh, *Jaegwan Dās Ki Bānī*, pp. 1-5; Crooke, *Tribes and Castes of the North-Western Provinces and Oudh*, Vol IV. pp. 299 ff; Prasadh, *Saṁtbānī Saṁgraha*, Vol. I. p. 117; Macauliffe, *The Sikh Religion*, Vol. I. pp. xlvii, xlviii; Grierson, *Modern Vernacular Literature of Hindustan*, p. 87.

Sāhib took from his pipe a blue thread, and Govind Sāhib from his a white thread, and these they bound on Jagjīwan's right wrist. (This is still the sign of the Satnāmī and is called *Amdū.*)

Jagjīwan then gave up all worldly cares and applied himself to study and devotion. Presently people began to come from a distance to see him. Thereupon a persecution arose, and he left his native village and took up his abode at Kotwa. The chief seat of his sect is still at this village, and here an annual fair is held. He was reported to have performed many miracles, one of the most famous of which was that connected with the marriage of his daughter to the son of Raja Gondā. When the Raja refused to partake of the wedding-feast unless flesh was served, Jagjīwan Dās created the egg-plant and this was eaten as meat. For this reason his followers still tabu that vegetable as convertible into flesh. He died in 1761.

Jagjīwan Dās preached the worship of God under the name "Sat Nām," and taught that the Deity is both cause and creator of all things, but conceived of him in popular Vedantic terms. His followers prohibit the use of meat, red dal (masūr kī dāl), egg-plant, and intoxicating liquors. Satnāmīs do not worship idols; but they do worship Hanuman, and pay reverence to what they consider manifestations of the nature of God visible in avatars, particularly in Rāma and in Krishna. They observe most of the Hindu festivals; and honour the family and caste customs of the members of their sect. Jagjīwan Dās urged that men should practise absolute indifference to the world, that they should be dependent upon no one, and that they should practise implicit obedience to the guru. They are said to practise the horrible rite of drinking a mixture of human secretions and excreta (gāyatrī kriyā). They enjoin tolerance, charity, consideration for others, prayer, study, and kindness to animals.

Jagjīwan Dās was, as noted above, a householder. The sect has a superior order of mahants, some of whom are of low-caste orgin. Through a Korī (weaver) disciple many Chamars and others of low caste were brought into the movement.

The principal works of this guru, written in Hindi, are the *Agh Bināś*, the *Gyān Prakāsh*, the *Mohāpralaya*, and the *Pratham Grantha*.

The Satnāmī movement was carried from Oudh into the Central Provinces by Ghāsi Dās, a Chamar, and there it has produced notable results. Ghāsi Dās carried on his great work during the decade 1820-1830. He was undoubtedly indebted to Jagjiwān Dās, for the teachings of the two men are well-nigh identical.

Ghāsi Dās was born in poverty at Girod, in the Central Provinces. As a man he took to the life of a pilgrim; later, he abandoned pilgrimage and began an ascetic life, and from that time retired to the forest regularly for meditation. The rocky hillock near his native village, to which he repaired, is still a place of pilgrimage. His reputation as a man of supernatural powers grew, and miracles were reported from his place of retirement. Finally, he emerged from the forest with his gospel to the Chamars. It was in substance the message of Jagjīwan Dās. "Ghāsi Dās, like the rest of his community, was unlettered. He was a man of unusually fair complexion and rather imposing appearance, sensitive, silent, given to seeing visions, and deeply resented of the harsh treatment of his brotherhood by the Hindus. He was well known to the whole community, having travelled much among them; and had the reputation of being exceptionally sagacious, and was universally respected."[1]

Ghāsi Dās died at the age of eighty years and was succeeded in office by his son Bālak Dās. The latter, however, managed things badly, and was assassinated in 1860. Since then the family has fallen upon evil times.

A division has occurred in the movement over the use of tobacco, and those who smoke use a leaf-chilam and not a huqqa.

Satnāmīs worship the Sun, morning and evening, as representing the deity, crying out, "Lord, protect us!" Otherwise they have no visible sign or representation of the Supreme Being. They are opposed to idolatry, and are

[1] Chisholm, *Bilaspur Settlement Report*, 1888, p. 45.

enjoined to cast all idols from their homes. Theoretically, they have no temples, no public religious service, no creed and no form of devotion. They simply call upon the name of God and ask his blessing. They, however, do have temples, and they recognize the whole Hindu pantheon, especially revering the Rāma and Krishna incarnations of Vishnu.

They profess to set aside caste and to receive all men as equals, but they do not admit into their community members of those castes which they regard as inferior to their own. The sect is practically a Chamar sub-caste. A Satnāmī is put out of caste if he is beaten by a man of another caste, however high, or if he is touched by a sweeper. Their women wear nose-rings, although Hindu law forbids it. They do not usually accept cooked food from the hands of others, whether Hindus or Mohammedans. With them two months are tabu for weddings, August (Shrāwan) and January (Pūs). An initiatory practice connected with marriage has already been described. It was carried out within three years of the wedding and after the birth of the first son.

The Satnāmī movement is of considerable importance as a social revolt on the part of the Chamars. As an economic and social struggle upwards it has met with a large measure of success. The history of the sect illustrates also how a theistic propaganda can live and transform a whole community.

There were, in 1911, 460,280 Satnāmīs in the Central Province, the number having increased about fifteen per cent. since the census of 1901.

CHAPTER IX

THE OUTLOOK

ONE of the outstanding facts about the Chamars is their lamentable and abject poverty. Ill-clad and cold in winter, badly housed, and insufficiently fed, they belong to the poorest of the land. While there are some well-to-do persons amongst them, and a few who are moderately rich, the great mass of the Chamars lead a wretched existence. Not more than one family in fifteen has any form of fixed tenure, and that only on *small* holdings. In many instances the hovels in which they live are repaired by the landlord, so that the Chamar may not acquire any claim upon the property. To begin with, they are greatly in debt on account of loans both for the purchase of raw materials with which to carry on their traditional occupation and for seed and for cattle for their agricultural enterprises. Rates of interest are exceedingly high, being from twenty-four to forty-eight per cent. on larger loans, and seventy-five on petty loans. In most cases their obligations are such as to keep them in perpetual bondage to their creditors; and as a consequence, they are never able to rise above the lowest economic level. In many instances the whole family is engaged in satisfying the insatiable demands of the zamindar or some other creditor. Many shoemakers in the neighbourhood of Delhi, for example, are so completely in the hands of the dealers of that city that they get but the barest living out of their hard toil. This economic condition suggests one important line of relief: the introduction of co-operative credit.

Another contributing cause to their poverty is the pernicious system of *begar*. Chamars live at the beck and call of others, and are obliged to do a great deal of work for which they receive no pay whatever. This is but a phase of the general condition of depression in which they live. They have been so conquered and broken by centuries of oppression that they have but little self-respect left and no ambition. Their condition is in reality serfdom, and at times they are sore oppressed. The begar system is firmly entrenched in the rural life of the country and can be broken down only by persistent and well-directed agitation. The old order must give way, even though the necessary substitute may be difficult to suggest. Their employers and the leaders of the Indian community bring social and even physical pressure upon them at times. Those who depend upon them for labour are slow to encourage any movement which brings to the Chamars an opportunity for advancement. They live on the land of others, and must bear without complaint oppression, injustice, and fraud. The solution of the problem which they present must lie in the bringing of economic help to them in the way of opportunity and encouragement, and in kindling in them a spirit of hope.

Another cause of their poverty is ignorance. Until their mental life is stimulated to the point where they begin to feel some sense of independence and desire for better things they will be held under the iron heel of those who exploit the poor.

A further cause of poverty is vice and excess. Intemperance is widespread. The Chamars are notorious drunkards and to drunkenness both men and women are addicted. Liquor has an important place in much of the domestic ritual. There is very little attempt to remove this evil. The only limit set upon it seems to be the income of the man or his family. *Ganja* (hemp for smoking), *charas* (hemp for smoking), *bhang* (hemp for drinking), and opium and *madak* (prepared from opium) are extensively used. Gambling is rife not only at Dewali time, but constantly. The Holi is an opportunity for excesses of all kinds. Children are not exempt from these evils.

One more contributing cause to the poverty of the Chamars is overcrowding on the land. For this reason a movement towards the large centres of industry must be encouraged. While it is important that improved industrial and agricultural methods be carried to the villages, it is also necessary that many be attracted to centres where instruction in industry and agriculture may be obtained. Such trained persons will very rarely return to take a place in the village economic life, but they will swell the ranks of organized industry and will help to reorganize agricultural and economic life. At the same time there will result increased demand for labour, and this in turn will raise wages and improve the general efficiency of those who are left to carry on agriculture. The movement to the cities has already set in and its effects upon the rural demand for unskilled labour are becoming more and more noticeable.

Seventy-eight per cent. of the Chamars are engaged in farm work. Here again they are found in the most wretched economic state. For the most part they are paid in kind, and there are few inducements offered to them to secure good crops. Moreover they are poor cultivators, and consequently obtain only the poorest portions of the land for farm purposes. While a considerable number will move to industrial centres, the great mass of Chamars will remain on the land. Advancement lies in their being taught better methods of agriculture. Moreover, these improved methods must be brought to them. They are far too numerous a caste to be sent into agricultural schools for training, and, besides, they could not be spared in any considerable numbers for such a purpose. Simple demonstrations that could not fail to convince the Chamars of the better economic values of modern methods of agriculture must be the means employed to introduce new methods.

Another cause of their poverty lies in the fact that the indigenous manufacture of leather is still in an undeveloped state as an industry, and that the output is of very inferior quality. For a long period before the Mohammedans began to rule, and even down to the present time, the rural industry has depended upon an inferior grade of raw materials, the skins of animals dying of disease or from starvation.

In addition to this, branding and injury ruin large numbers of hides. At present, with the marked rise in values, hides and skins, except those of the very poorest quality, are becoming more and more difficult for the Chamār to obtain. Added to this, lack of capital makes it impossible for the village tanner to make good leather. Good tanning requires time and that in turn requires money. The result is that the rural tanner, using antiquated and inferior methods, produces out of poor raw materials a very inferior grade of leather. The outstanding defects in the village process are over-liming, the use of antiquated tools for fleshing and removing the hair, insufficient attention to bating, the hurrying of the process of tanning, and little attempt at currying. With the rise of the large-scale tanning industry in certain large centres, the village tanner's enterprise is being reduced to smaller dimensions. There is little likelihood that the rural industry will survive.[1] In this connection it is interesting to note that during the decade ending in 1911 there was a very marked decrease (36.9 per cent.[2]) in the number engaged in tanning, currying, dressing, and dyeing leather. At the same time the Chamar population increased. Furthermore one of the results of the war has been a very great advance in large-scale tanning. The demand for village tanned leather is gradually being reduced to that of water-buckets and thongs. The former will be supplied more and more from chrome tanned leather, which is not a rural product at all, and finally, cheaper fabrics made from vegetable fibres will supplant leather for irrigation purposes. Slowly factory tanned leather will supplant village tanned leather in the village shoemaking industry.

Before the war raw hides were exported from the United Provinces in large numbers.[3] In 1914-15, the exports of dressed or tanned skins amounted to only fifteen thousand rupees; while that of raw hides and skins

[1] *Indian Industrial Commission Report*, 1918, p. 36.
[2] *Census Report, United Provinces*, 1911, p. 424.
[3] *Report of the Director of Industries, United Provinces*, 1916, p. 4

amounted to 1,84,50,000 rupees. Probably half the hides and nine-tenths of the skins available in India were exported. Not only have those exports risen to enormous proportions in recent years, but the values have likewise increased. The total value of exported raw hides and skins was 7,82,00,000 rupees in 1914-15 and 14,41,00,000 rupees in 1916-17. During the same years the values of exported leather and of tanned hides and skins were 4,76,00,000 and 9,44,00,000 rupees respectively. The total values of these exports were 12,58,00,000 and 23,85,00,000 rupees in 1914-15 and in 1916-17 respectively.[1]

During the war the amount of half-tanned leather exported from the United Provinces increased from below 200,000 hundredweight, valued at less than 2,00,00,000 of rupees to 360,000 hundredweight, valued at nearly 5,00,00,000 rupees in 1917-18. Roughly speaking, in four years the output of the Indian tanneries for this class of leather only has been doubled.[2] In all probability the enormous demand for hides and leather due to the effects of the war on stocks of cattle in Europe will turn to India's advantage. With the development of tanning materials and the application of technical skill and expert direction to the manufacture of leather in India, there will be a large increase in the tanning industry in manufacturing centres. For this new development the Chamar is indispensable. But this new stimulus to enterprise will tend to further supplant the village tanner. The development of this industry involves the training of large numbers of Chamars. This suggests one of the lines along which work for the economic uplift of the Chamars must develop.

There were in 1911, in the United Provinces, all at Cawnpore, three tanneries and ten leather factories. Not one of the latter was managed by a Chamar.[3]

While the number engaged in the tanning of leather decreased very materially during the decade ending in

[1] *Appendix D, Indian Industrial Commission*, p. 54.
[2] *Appendix D*, p. 58.
[3] *Census Tables, United Provinces*, 1911, pp. 720, 736.

1911, there was an increase of 33.2 per cent.[1] in the number engaged in the manufacture of boots, shoes and sandals. The Indian demand for boots, shoes and sandals is on the increase, and this phase of the Chamar's traditional occupation offers increasing opportunities. At present the native patterns of ornamented shoes are disappearing and shoes on foreign models are coming largely into vogue. The great cities are the centres of this industry. But shoes after the country models are manufactured in almost every village in the country. Here also there is great need for the introduction of better tools and modern methods of manufacture. And a growing field for demonstration work and industrial education here presents itself.

There were in 1911, in the United Provinces, four boot and shoe factories, one in Allahabad, two in Cawnpore, and one in Farukkhabad, not one of which was owned or managed by Chamars or Mochis.[2]

The demand for other kinds of leather articles gives promise of still further developments in the leather industry. Belting, roller skins, picker bands, and raw hide pickers will be required in increasing numbers with the rapid industrial development of the country. Already a beginning has been made in supplying these products in India. There seems to be little doubt that, now that the war is over, new tanneries will be started, and their fate will largely depend upon the quality of the leather which they turn out. Here Government can render valuable assistance by assuming to a large extent responsibility for the techno logical investigations which have been indicated. Success will result in an improvement of the industry all along the line, beginning with a decrease of waste in rural areas and the diversion of the hides used by the village tanners to modern tanneries, in which a better class of leather will be produced. There will obviously be an increase in the amount of visible raw material; but whether this will be sufficient to meet the growing requirements of the country

[1] *Census Report, United Provinces*, 1911, p. 425.
[2] *Census Tables, United Provinces*, 1911, pp. 728, 740.

is a matter on which no definite opinion can be expressed. The general improvement of the technique in tanning will lead to an increase in the exports of finished leather and to a corresponding decrease in the exports of hides.[1]

These conclusions suggest, for those who are especially interested in the Chamars, that they may take advantage of facilities offered through Government for the training of men in various phases of work in leather.

In all industrial development there must be a safe-guarding of the Chamars' interests. This must be done through legislation which will protect the Chamars from overcrowding in the growing manufacturing cities, and by framing laws fixing reasonable hours and liberal wages for the labourers. Such legislation is dependent upon organized effort on the part of those who champion the cause of the Chamars.

Education amongst the Chamars is exceedingly backward. Below is a table of literacy based upon the Census Report of the United Provinces for 1911.[2] Along with the figures for the Chamars those from the population of the Provinces as a whole are inserted.

Number of persons literate per thousand :

		Total	Male	Female
(1) Chamar (agricultural)	..	1	2	(.2)
(2) Population as a whole[3]	..	34	61	5

The number of Chamar children in primary schools per thousand males is .3, and per ten thousand females is .1. In 1917, there were 4,600 Chamars undergoing education in the United Provinces.[4] These tables give a very inadequate impression of the ignorance that prevails. A more detailed statement showing both the literates and the illiterates in this caste is as follows :[5]

Chamar pop. dealt with			Literate	Illiterate	Literate in English
Total	6,068,382	..	6,794	6,061,588	215
Male	3,099,321	..	6,274	3,093,074	214
Female	2,969,061	..	520	2,968,541	1

[1] *Appendix D., Indian Industrial Commission*, pp. 64, 65.
[2] p. 273. [3] p. 268.
[4] *General Report on Public Instruction in the United Provinces and Oudh for the Quinquennium Ending 31st March, 1917*, p. 94.
[5] *Census Tables, United Provinces*, 1911, Table IX., p. 154.

A still more definite impression is given by another table:

The illiterates per thousand are:[1]

	Total	Male	Female
Chamars (agricultural)	999	998	(1,000)

Ignorance is more deeply seated than the mere inability to read or write. For generations Chamars have been, and they still are, out of touch with even the best light that the village possesses and their mind is almost inert. These conditions are but barely improved in some areas where education has long been emphasized.

The public schools are virtually closed to the Chamars. Both teachers and pupils in the schools make it most difficult for low-caste boys to sit in the class-rooms. The result is that boys of the lower castes are not found in any numbers in the schools. A typical case may be cited. In a school enrolment of 12,651 in a certain District there was, in 1909-10, not a single Chamar.[2] Conditions are not much changed yet. Schools are now being established for the class to which the Chamars belong, and Government is encouraging the opening of such schools by District Boards. Besides this, efforts are being made to set aside special educational officers for schools for the depressed classes. Moreover, various Hindu organizations are trying to carry on primary schools amongst these classes. Still, up to the present time practically the only opportunities for learning to read and write are supplied by Christian agencies.

Besides the lack of educational facilities and their intellectual inertness, the environment in which the Chamars live is unfavourable to their advancement. Their neighbours, who largely control their time, are not interested in enabling them to attend school with any regularity. The feeling is widespread that an ignorant Chamar is the only useful Chamar. Enlightenment in the least degree brings with it (so it is held) a certain

[1] *Census Report, United Provinces*, 1911, p. 273.

[2] Bijore District, United Provinces, Letter from Deputy Inspector of Schools, March 18, 1911.

sense of personal importance and the desire to be free; and all this is contrary to the spirit of their environment. The intellectual uplift of this great caste is a tremendous problem, but one of the greatest importance in the advancement of the whole country. Every eighth man in the United Provinces is a Chamar. This fact illustrates both the weight of the ignorance that oppresses the land and the possibilities for social and political advancement which lie in the uplift of this depressed group. In this day of emphasis and expansion in primary education the Chamars offer one of the most wide and needy fields for cultivation.

Most important is the question, "What shall the Chamars be taught?" Of course they must learn to read, write and cipher. This must be accomplished through day schools conducted at such times as pupils can be spared from their regular tasks, and by means of night schools for adults. But it is equally important that they receive instruction that will open their minds to moral and religious truth that has in it the power to emancipate them from superstition and fear and the spirit of servitude. Furthermore, since an effective intellectual and religious development cannot be based upon poverty, the educational programme must include instruction in improved methods of industry and agriculture. And since the men who go away from their villages to learn something new rarely return to join in the village industrial or agricultural life, such training must be brought to them in their village environment. Improved methods of tanning, of making shoes, of weaving, and of cultivation must be brought to them by means of demonstration work. Already Government is busy with plans for these kinds of simple instruction and is also applying these methods in a few places. Such forms of instruction must become part of the curriculum in all schools which aim at the elevation of the Chamars; and the agencies which will develop with vigour and foresight such forms of educational endeavour will have the greatest degree of access to the caste.

The Chamar holds a place very low in the social scale. He belongs to the "untouchables." This is due partly to ignorance, more to his poverty, and still more to his being a subject caste. The long history of conquest may be read here; and here also the fact that those who are oppressed are always despised is amply proved.

But the sense of disgust which he arouses is due also to his traditional occupation. His name associates him with dead animals. But to the ordinary Indian a dead animal suggests not only a skinner, but also a group of Chamars, men and women, dividing and portioning out the carcass and preparing for a feast.

Furthermore, they eat the leavings of food of most castes. This also is an abomination.

Added to all this is the unclean condition of the places where they live. Their tanning vats are just outside of their houses, and their part of the village is a place of all sorts of abominable smells. Sanitary laws are wholly ignored. They are unspeakably filthy in their habits. Their persons, their clothing, their houses and their surroundings are utterly unclean. The *Chamrauṭi* is a synonym of all that is unclean and disgusting. A further abomination is the fact that the Chamari is the recognized midwife of the community (with local exceptions). The word Chamari is sometimes used as a synonym for midwife. Her offices are considered as exceedingly polluting. So the Chamar's quarter of the village is a place to be avoided, and Chamars are too unclean to enjoy any of the social or religious privileges of the Hindu community. Even in bathing in the Ganges they must find a place far below that used by other people. In Madras the leather worker pollutes at a distance of twenty-four feet. Conditions are much less rigid in the North.

But skinners and tanners find themselves by reason of the nature of their work in a very low social position, and while the conquered have had to find their living among the despised, still, there are other elements that have helped to confirm these low-caste groups in their social positions. The idea of pollution, or its reverse, the idea of purity, may be traced more accurately to worship. The sense of ceremonial purity certainly antedates the idea of pollution due to the eating of beef or to the idea of the sanctity of the cow. It was the right to share in the fire-sacrifice that was early restricted. When the worship of the cow came into vogue, the idea of pollution was intensified. The sense of separation once made absolute on the grounds of ceremonial pollution, the whole life of the group, habits and occupation included, were taken up into the attitude of disgust. It was thus through religious scruples that the racial element was joined with the occupational to fix the social level of such as the leather-worker. The men with the disgusting occuption were of an alien race and religion, and by that very fact impure. If any further considerations were necessary to complete the realization that the leather-workers were outcastes, it would be found in their affinities with non-Aryan races in matters of belief. There is much in their superstitions and in their customs, and there always has been, that sets them off by themselves so far as the Aryan or the Hindu is concerned. In this worship there is at least an expression of the sense of some superior power, though that power is most often malevolent, and the accompanying sacrifice is to appease or to propitiate the object of the voiced entreaty or request. The whole range of primitive praying, from the worship of the fetish and the totem to the adoration of the scarcely-known higher gods, is present in the religious life of the Chamar. But, by the side of this personal, social element, there is the anti-social, anti-religious use of charms and spells which belong to magic. The Chamars have a reputation for witchcraft, and this is borne out by abundant practice both of white and black magic. Again, while the domestic ceremonies of the Chamars show much Brahmanical influence, and while the cardinal elements of Brahmanism are practised by them, still there is a very large admixture of details of ritual that belong to the non-Aryan religion. The fear of demons and the principles of spirit-possession are everywhere taken into account, and malicious spirits and demons of disease are universally feared. None of these elements of primitive belief are borrowed; they come from the strata in which the Chamars themselves are found. These facts also set Chamars at a tremendous social disadvantage.

To the foregoing reasons why the Chamars are despised above most men may be added the reputation which they have for crime. They are popularly regarded as poisoners of cattle. In the Chhattisgarh Division of the Central Provinces they are regarded as the most criminal class in the community. Their reputation for crime is undoubtedly far beyond the facts. All of these factors which combine to fix the Chamar's social status bear testimony also to his social condition. Another social fact is the laxity which exists in matters of morality. While some forms of adultery are severely dealt with, there is much impurity, and the general thought-level in matters concerning the relations of the sexes is very low. The *nach*, in which men and boys dress as women, and in which women take part, is another evil. Obscene songs and coarse jesting are very common. Women are held in very little respect. The picture of the social aspects of the Chamar's life may be completed by reference to the state of education in the caste and to religious beliefs and social customs.

There is difference of opinion concerning the physical fitness of the Chamars. Poverty, intemperance, and lax social standards, together with the practice of child-marriage, would naturally combine to make them men of inferior physique, and yet some think that they are strong men capable of great endurance. The judgment that is passed upon the Chamars in this respect depends very largely upon the locality which the judge has in mind.

However, so far as infirmities are concerned, the Chamars compare favourably with the population as a whole. The figures on infirmities among the Chamars for every one hundred thousand of the population of the United Provinces in 1911 were as follows:

Insane		Deaf-Mutes		Blind		Leper	
Male	Female	Male	Female	Male	Female	Male	Female
17	9	50	36	208	288	43	12
23	12	67	45	209	236	48	11

Only in the case of blindness are the afflictions more numerous amongst the women. The corresponding number for the whole population of the United

Provinces per one hundred thousand are given for comparison.[1]

The outstanding fact about the Chamar's religion is its lack of comfort and of hopeful outlook. For the most part he lives in fear of malevolent powers and is engaged in propitiating them, and superstition grips him with all its terrors. Furthermore, the great mass of the Chamars know very little about their own religious beliefs and social customs. "This is our custom"; or, "This is the way it is done," is their usual answer to questions. For example, being questioned, a village Chamar replied that he was a "*Pachpiriya.*" But he could not name a single one of the "Five Saints" whom he worshipped, nor could he give any information about his religion. All that he could say was, "I am a *Pachpiriya.*"

This ignorance concerning their religion leads many of them to say that they have no religion at all. Although there are considerable numbers of Chamars who follow the gurus of the reform sects and who have risen through initiation to a relatively higher religious plane, the religious and moral conceptions of the masses of the Chamars are reflected in the domestic customs and in the attention paid to demons of various kinds. The domestic customs contain mainly three elements: (1) obscenity and intemperance, (2) superstition, and (3) idolatry. Where the Chamars have lived for some time in the larger cities, and where they have come under the influence of the Arya Samaj, or of Christianity, they are becoming ashamed of the grosser and more superstitious elements in their customs, and are professing to have lost faith in their godlings.

The things for which they pray are mostly of the material sort, since they have little hope of obtaining spiritual benefits from those to whom they address the longings of their hearts. They ask, for the most part, to be let alone, not to be plagued nor to be overtaken by calamity. Religion consists in doing (*karam*), in performing

[1] These figures are approximate and are based upon *The Census Report, United Provinces*, 1911, Subsidiary Table I., pp. 320, 317, 318.

customary acts, in bathing, in making offerings, in pilgrimages, and in similar Hindu practices. They are anxious to fulfil the Hindu requisites of life. The future holds no great prospect for them, for they are very low down in the transmigratory world. Only as good works may modify the possibilities of the future are they seriously concerned with duty. For them the chief end in life is to live as comfortably as possible, to obtain the largest possible share of "pleasure," and to escape as many of the untoward experiences of life as possible. Of the violation of the moral law they have some notions; and they agree that it is good to be honest, truthful, chaste, kind, generous, and hospitable; but, in this hard world, such standards of life are difficult to attain; consequently, Chamars are not over zealous in good works. They admit that such works are good for those who do them. Still, there has been widespread religious advance, coupled with insight and enthusiasm, with the acceptance of the message of *bhakti*, or devotion, through the theistic reform sects. This is especially noteworthy in the movements issuing from Rāmānanda. In this phenomenon there is ample assurance that the Chamars may have a much better future.

The response of the Chamar to the influences of the great socio-religious forces about him is marked. First, there is the general steady effort to follow orthodox Hindu customs. Caste fissures also bear testimony to the influence of Hinduism.

Second, there is the response to the efforts of the Arya Samaj. The last Census (1911) recorded 1,551 Arya Chamars in the United Provinces.[1] In some areas considerable effort has been made by this organization. But up to the present time they have not formulated any broad policy. There are isolated efforts, however, and a broader policy is sure to appear. As yet the Arya Samaj confines itself largely to those communities where other religious bodies have already begun to work, and enters these places to a considerable extent as an obstructionist.

[1] *Census Tables, United Provinces*, 1911, p. 301.

Third, the influence of Islam is marked. The Julahas, who are Mohammadans, and the Mochis, most of whom are Mussulmans, are standing witnesses to the influence of Mohammadanism. Besides this, 5,651 Mussulman Chamars were reported in 1911 in the United Provinces, and 10,811 in all India.[1]

During 1911, preceding the Census enumeration, both the Arya Samaj and the Mohammadan communites made special efforts to enrol Chamars, especially those who were Christians. With the increased rivalry between the two communities, as representative government gains ground, both will make greater efforts to win the Chamars.

Fourth, the Christian Church is gaining a good many converts from amongst the Chamars. Christian converts are being made in a number of widely scattered areas, and so-called mass movements amongst leather-workers are now in progress from the far South to the North of India. In Northern India the largest movements are in the liberal-minded areas in the north-west of the United Provinces and in Bihar. At the present time fully half a million Chamars are being directly influenced by Christian propaganda, and many thousands more indirectly. The knowledge of this movement is very widespread amongst the Chamars of all sub-castes. Reports from many areas indicate that as a caste they are accessible. Already some 45,000 have been baptized.[2]

The problems confronting those who undertake to lift up the Chamars are very great. Neither upon ignorance nor upon poverty can any large advance be made. A real programme for their economic uplift is in itself a very large task. Co-operative societies, improved methods in industrial and agricultural work, and the emancipation of the Chamar from the thraldom of begar are involved in this problem. And the very great size of the caste makes the problem still more difficult. No large educational advance can be

[1] *Census Tables, United Provinces*, 1911, p. 278; *Imperial Tables*, 1911, p. 187.

[2] The estimates given above are conservative, and are based upon incomplete returns from missionaries.

expected until there is a real improvement in their economic conditions. Much of their lethargy, much of their indifference shown towards education, is due to the depressing influences of poverty. In addition, hygiene, sanitation, and domestic economy must find their place amongst the masses through the development of real social centres. We thus have a large field for a real social programme. Added to this must be a moral and religious programme which will stamp out drunkenness and immorality; which will give them a real sense of personal worth and a feeling of self-respect; and which will alienate them from superstition. Then the outstanding elements in domestic rites and customs, obscenity and vulgarity, superstition and idolatry, must be eradicated. Enlightenment and moral teaching will deal most effectively with the two former, and pure religion with all three, and especially with the last. The process of emancipation will not be rapid. Their case calls also for regulative laws, for a legislative programme. One outstanding need is that of the planting of Christian social centres in Chamar communities. Community organizations, of which the settlement house and children's houses are suggestive, are here needed.[1] In the village life of this great caste is found one of the greatest opportunities for social endeavour such as that urged by the foremost Christian leadership of to-day. This will mean that the Chamar is offered a real

[1] The Board of Home Missions of the Presbyterian Church in America operates a Department of Immigrant and City Work. In ministering to the recent immigrant the approach is from the community standpoint, *i.e.*, the immigrant is dealt with, not detached from, but in relation to, his environing community. Knit up with the community in which he lives is his life, his progress, his welfare, and that of his family. Therefore the work of the Church in any given immigrant community must be developed on a comprehensive scale. "It calls for a sympathetic understanding of the previous life and social and religious traditions of the immigrant, and at the same time demands that we introduce him to the best this country has in civic, social and religious ideals." This principle has inspired a new form of religious ministry conducted by the Presbyterian Home Board. It is designated "industrial parish work." This work is now operating in no less than nine important industrial communities where the new immigrant is a large population factor.

social fellowship[1]; that his economic and educational claims will receive attention; and that in work for his emancipation the most modern methods of religious education will be introduced. Those who would help him must sympathize with him in his superstition and his degradation and go to him with positive teaching rather than with attack upon his customs and beliefs. The elements of prayer and of belief in spiritual values are potential in his fears and superstitions.

In the foregoing paragraphs some social and religious problems of the Chamars have been discussed, and some solutions have been outlined. There remains to suggest some means through which the Chamars may be lifted to a satisfactory place in the social order and through which they may enter into a satisfying religious life.

In a real sense the Chamars are the product of the social and religious teachings of their own land. According to the doctrine of karma a man is what he is because of what has happened, and he finds himself just where, in the very nature of things, he belongs. Chamars, and their neighbours in the social scale, are foreordained to menial tasks with no outlook towards better things *in this life*. Ignorance, grinding poverty, servitude and degradation *are* their lot, and, although there are many signs of a new day for these "untouchable" classes, still movements urging improved conditions for these outcastes, which are now stirring in many parts of India, arise from impact with Christianity, rather than from the social force of Hinduism. Says a noted Indian, "The ideas that lie at the heart of the Gospel of Christ are slowly but surely permeating every part of Hindu thought."[2] While the religious teachers of India do not present an adequate social programme for the Chamar,

[1] This is a conception hard to be grasped in a country with a social history like India has. In a recent discussion of social service the following sentence is found: "The soulless animal rises up at the command of the teacher metamorphosed into a full-fledged human being. . ." *Indian Review*, May 1918, Article "Social Service in the Punjab."

[2] Farquhar, *Modern Religious Movements in India*, p. 445.

Jesus does.[1] The Law and the Prophets which he came to fulfil champion the cause of the weak and condemn those who exploit the defenceless and the poor. The striking thing about Jesus's message is his estimate of the common people, the peasants and the common labourers. "Ye are the salt of the earth"[2] follows after:

> "Happy are you poor !
> For yours is the Kingdom of God.
> Happy are you who hunger now !
> For you shall be satisfied."[3]

Here is the message of economic salvation. Jesus insists that men are not impersonal units to be herded together, or to be exploited, or to be sacrificed to the whim of the more fortunate classes, but that they are valuable persons to be delivered from their hard lot. He is the champion of the depressed masses. His message has always been good news to them. A very large part of the growing church of the first century was made up of "Wool-dressers, cobblers and fullers, the most uneducated and vulgar persons."[4] Since then new life with new privileges and new living conditions has followed in the wake of the acceptance of Christ. And to-day we find that amongst the poor the leaven of his economic promises is at work. The exploited and the poor are voicing demands which are big with expectations which Jesus has encouraged. And the Chamars are beginning to look hopefully to Christianity for emancipation.

Again, the Chamars are by birth doomed to illiteracy. Indian traditional ideals concerning the privileges of enlightenment are well known. On the other hand, the prophets and Jesus were *teachers.* The classes whom they championed, as well as those who sat in authority, were their pupils. The new Christian, even from the

[1] For a clear analysis of this subject see Kent, *The Social Teachings of the Prophets and Jesus.*

[2] Matthew v. 13.

[3] Luke vi. 20, 21.

[4] See Glover, *The Conflict of Religions in the Early Roman Empire*, p. 241. This book makes a good commentary on the subject of these paragraphs.

despised classes, became a teacher.[1] The children of Christian parents were taught. The Christians acquired knowledge. And it has been the pride of Christianity from the beginning that it has developed an enlightened community. The advance of popular education in the West has been *pari passu* with an open gospel.[2] So far, in India, the only real heralds of enlightenment for the Chamars have been Christians. No other agencies are yet able to place open books and liberty of thought before all men equally.

Besides, Jesus has a programme for the socially disenfranchised. It rests in the recognition of the solidarity of the race. He draws no artificial lines of division in society. Jesus gave practical illustration to his principles. He mingled freely with all classes. He accepted with equal alacrity invitations to dine whether given by learned Pharisees or by despised tax-collectors. When Scribes and Pharisees flung at him the contemptuous charge that he was the friend of drunkards and social outcasts, Jesus openly declared that the men who appealed most strongly to his sympathies were the socially disinherited classes; those who were ceremonially and morally beyond the pale of Pharisaic teaching; those who were regarded by the religious classes as little more than social refuse. Most of them were social outcasts, and many of them, because of their crimes and manner of living, were probably debarred from the synagogues. Such facts as these show how fully the Chamars' need for a new place in the social order is met by the position of Jesus.

Moreover, there are signs that Chamars are open to such a gospel as Jesus preached. They are uneasy; they dislike being called Chamars; they are anxious to shake off the disgusting practices connected with their name; they long for a better place in society; and they desire economic freedom. They are beginning to look towards Jesus for the realization of these things. Indeed, in many

[1] Glover, *The Conflict of Religions in the Early Roman Empire*, 241, 242.

[2] Simple illustration of this point may be found in any good history of the English Bible.

places the more enlightened amongst them feel that real opportunities lie with the Christ; and they are willing to endure persecution and many hardships for the sake of the gospel. From widely separated areas this attitude is reported. Beyond these groups there are many others who say that they will all eventually enter the Kingdom of God.

Jesus stands in relation to the Kingdom of God as the Saviour. He is the initiator of the new order. He is the deliverer of those who are bound. He indentifies himself with the hungry, the naked and the lowly, and he says that he came to save them. The Kingdom comes after he has paid the price. His death puts the seal of sincerity upon his words. Not even the pangs of death could make him yield his position. Thus the powers of evil are forced back and men enter into a new life.

Their question of social purity must be studied from the standpoint of Christianity. Standards of sex relationship are set forth once and for all, and in final terms, by Jesus, who speaks with supreme insistence:

"But I say unto you, that every one that looketh on a woman to lust after her hath committed adultery with her already in his heart."

"And if thy eye causeth thee to stumble, pluck it out, and cast it from thee: for it is profitable for thee that one of thy members should perish, and not that thy whole body be cast into hell."

"And if thy right hand causeth thee to stumble, cut it off, and cast it from thee: for it is profitable for thee that one of thy members should perish, and not thy whole body go into hell."[1]

Obscenity and laxity cannot live under such surgery. Jesus says further:

"Whosoever shall divorce his wife in order to marry another, commits adultery against her." He puts the other side of the case also: "And if she divorces her husband in order to marry another, she commits adultery."[2]

[1] Matthew v. 28, 29, 30.

[2] See Kent, *The Social Teachings of the Prophets and Jesus*, p. 243.

Jesus bases his teaching concerning adultery and social impurity on the worth of the individual. Acts of sexual immorality are traced to the failure to appreciate this point of view. For Jesus the family is the fundamental institution, and in it woman finds her true place of worth. Thus he puts a supreme value upon personality. On the other hand Hinduism seeks and guarantees the absolute loss of personal identity of character, of consciousness, and of all else which makes for a glorified humanity. The doctrine of *maya* takes all worth out of personality. Besides, the social conditions as they exist in the lower strata of society are taken for granted; they are simply the working out of inevitable law (karma). Moreover, the Indian Law Books look upon woman with suspicion. On the foundations of such a social conscience salvation in its deeper social significance can never be achieved. But Jesus, on the other hand, draws men out of the pit and establishes them in clean living.

The history of the early church gives testimony to the power of Jesus to purify life. It is never to be forgotten that the great numbers in those days were from the "lower" ranks of life. Yet "they were astonishingly upright, pure and honest," and "they had in themselves inexplicable resources of moral force."[1] "The early Christian rose quickly to a sense of the value of woman."[2] This is all the more remarkable when the low moral standards in those Roman days are taken into account. But such purifying of life has taken place in every age and country where Jesus has been accepted. The Chamars need just such a Saviour as Jesus has proved himself to be.

Christianity is however for the Chamars more than an economic, more than a social gospel. Hinduism on its lower side is polytheistic, is saturated with demonology, and is exceedingly superstitious. Therefore the Chamars must look elsewhere for deliverance from superstitions and the fear of evil spitits and from the evils which follow in

[1] Glover, *The Conflict of Religions in the Early Roman Empire*, p. 142.

[2] *Ibid.*, p. 163.

the wake of these beliefs. Historically Christianity is the religion which emancipates from such forms of bondage. Jesus has always cast out devils. Christianity from the first has witnessed the disappearance of pagan beliefs and fears. A glance at early Christian history shows this. In the first century people hung rags and other offerings on holy trees, revered wells and streams, and believed in magic, enchantment, miracles, astrology and witchcraft. They were in the grip of demons with their hatred of men, their immorality and cruelty, and their sacrifices, and they knew the terrors of "possession" and of enchantment. Christianity came as a deliverer, and in the place of terror came peace. A new phenomenon, Christian happiness, or sense of security, appeared. Since then Jesus has driven back the forces of darkness in land after land. Down to the present time he rebukes superstition with the same authority and gives peace to fearful souls. In India to-day there are thousands of followers of Jesus who scarcely know the names of the demons whom their parents feared.

Besides, Jesus has a message about God. According to him the poor as well as the rich will find a Father. Jesus's teaching is about an active, sympathetic, sufficient Person. God is not lost in the shadows nor set afar off by lesser beings. He is close at hand and on the side of the poor. Moreover, God is not the great Terror. Jesus revealed him as the great Father. Nor is God a mighty Despot sitting high over his subjects; he is a Father who forgives sins freely, welcomes the prodigal, makes his sun shine on the just and the unjust, and who asks for nothing but love, trust, co-operation and obedience. Such a God will attract the mind and heart of the Chamars. Their hope lies in the realization of Jesus's teaching of the Fatherhood of God, and of its corollary the brotherhood of man.

Furthermore, Jesus himself offers the sufficient life for men. He did not fail where others have failed.[1] With his unique ideas about God and his intimate fellowship with

[1] See W. Rauschenbusch, *A Theology for the Social Gospel*, pp. 155-159.

God, he did not lose touch with the hard realities which confront men everywhere. While he is our great example of prayer and immediate communion with God, still the Kingdom of God engaged his will and set his task in the midst of men. He drew his strength from God, but he put it forth in the world. The needy world drew forth all his sympathies. He was not a mystic in the narrower sense of escape from the world. Furthermore, Jesus did not read life in terms of its darker side. He was not a pessimist. Although he knew the strength of the Kingdom of Evil, he launched the Kingdom of God against it and staked his life on the issue. Even when his life was overshadowed by opposition, seeming failure and death, he did not despair. Besides, Jesus was neither ascetic nor other-worldly. He liked a normal well-rounded life. He set forth the distinctive difference between himself and John the Baptist, showing that he placed himself in the midst of men in their everyday life and needs. He believed in a life after death, but it was not the dominant element in his teaching nor the constraining force in his religious life. He was concerned with the well-being of men in this world. He fasted when he was absorbed in thought. He went without food, sleep, and home life because he was set on a big thing. This is the revolutionary asceticism of the Kingdom of God, but that is wholly different from individualistic and other-worldly asceticism. Jesus communed with God; he fully recognized the power of evil in the world, and he held his life with a light grasp. Yet he escaped the snares of mysticism, pessimism, asceticism and other-worldliness. Out of the same ingredients, communion with God, recognition of evil, and religious intensity and self-control, he built his higher, sufficient life. His attitude toward life was the direct product of his two-fold belief in the Father who is love and the Kingdom of God which is coming.

And finally, Jesus is the sufficent object of devotion. The announcement of Rāmānuja to the low-caste man was that he can worship God, and that he has a real religious nature. In Jesus we find the fullest scope for the life of devotion. Jesus draws all right thinking and feeling, all high motives, all clean hearts, all new-made men and women, all devotion and love of all men up into his great and sufficient life. He is the perfect ideal of life and love.

Thus Jesus offers to the Chamars a satisfactory place in the social order and a satisfactory religious life.

Those who hold that in Christianity lies the real hope for the redemption of the Chamars are confronted with the fact of the urgency of the problem. Unless the Christian Church pushes forward with a broad programme, opposing religious movements may gain advantages which it will be difficult to surpass. In the end, however, only that movement can succeed which is able to give the Chamars, be it ever so slowly, character, the ideal and the reality of good citizenship and a satisfying religious experience. And although the task may look very large, many are confident that the full redemption of the Chamars will come through the Gospel of the Son of God. The forces that confront pure religion in the beliefs and superstitions of the Chamars to-day are not unlike those that opposed Christianity in the Roman Empire. Gibbon wrote: "The innumerable deities and rites of polytheism were closely interwoven with every circumstance of business and pleasure, of public or private life, and it seemed impossible to escape the observance of them without, at the same time, renouncing the commerce of mankind and all the offices and amusements of society." Yet the old order passed away. The Christian Church looks for no less a victory for the Chamars. Its justification for its superhuman undertaking is, "The poor have the Gospel preached to them."

APPENDIX A—(TABLES)

(A) DISTRIBUTION OF CHAMARS, MOCHIS, JULAHAS AND KORIS BY RELIGION AND LOCALITY[1]

	Chamar (Chambhar)[2]	Mochis	Julaha[3]	Kori
Persons ..	11,493,733	1,018,366	2,858,399	918,820
Hindu	11,305,713	561,777	98,651	917,633
Sikh	175,379	195	6,233	25
Jain	18	..	..	..
Buddhist	114	..	..	..
Mussulman	10,811	456,120	2,763,473	1,162
Animist	1,696	274	..	..
Ajmer Merwara ..	13,351	..	..	..
Assam	54,234	13,697	7,207	..
Bengal	136,553	455,448	282,705	..
Bihar and Orissa ..	1,114,467	31,339	826,391	..
Bombay	302,536	:.	10,478	..
Central Provinces and Berar	907,927	4,007[4]	..	39,628
Madras	..	6,285	..	..
N.-W. F. Province..	..	23,209	37,390	..
Punjab	1,139,941	419,378	635,044	18,050
United Provinces..	**6,083,401**	**8,869**	**991,263**	**860,434**
Baroda State ..	32,210	8,954	..	..
Central India Agency	859,438	7,126	12,270	..
Hyderabad State ..	70,618	7,023	5,194	..
Kashmir State ..	39,099	14,694	26,830	..
Rajputana Agency..	734,110	15,428	18,333	..
Elsewhere	5,848	6,916	5,294	708

[1] Based upon *Census of India*, 1911, Vol. I. Pt. II. Table XIII. Pt. I.

[2] Includes Dabgar in Bihar and Orissa and Mochi, Mochigar or Sochi in Bombay. The Sochi of Sind, however, has been shown separately. (*Census of India*, 1911, Table XIII. pp. 187, 213.)

[3] Hind Julaha in Bengal and Bihar and Orissa not included.

[4] Not in Table.

(B) Chamars in relation to the Whole Population of the United Provinces

Name of District	Total Population in 1911	Chamar Population in 1911	Per cent. of Population in 1911 that was Chamar
1. Dehra Dun	205,075	24,496	11+
2. Saharanpur	986,359	202,268	20½
3. Muzaffarnagar	808,360	124,459	15+
4. Meerut	1,519,364	236,834	15½
5. Bulandshahr	1,123,792	184,985	16+
6. Aligarh	1,165,680	194,013	17—
Total of Meerut Div.	5,808,630	967,055	17—
7. Muttra	656,310	102,757	16—
8. Agra	1,021,847	174,006	17
9. Farrukhabad	900,022	98,891	11—
10. Mainpuri	797,624	105,719	13+
11. Etawah	760,121	119,086	16—
12. Etah	871,997	115,382	13—
Total of Agra Div.	5,007,921	715,841	14+
13. Bareilly	1,094,663	98,506	9—
14. Bijnore	806,202	136,544	17—
15. Badaun	1,053,328	148,032	14+
16. Moradabad	1,262,933	180,957	14+
17. Shahjahanpur	945,775	100,061	10+
18. Pilibhit	487,617	34,005	7—
Total of Rohilkhand Div.	5,650,518	698,105	12+
19. Cawnpore	1,142,286	138,075	12
20. Fatehpur	676,939	72,553	11—
21. Banda	675,237	106,330	16—
22. Hamirpur	465,223	69,263	15—
23. Allahabad	1,467,136	162,735	11
24. Jhansi	680,688	92,357	14—
25. Jalaun	404,775	68,762	17—
Total of Allahabad Div.	5,494,284	710,075	13—
26. Benares	897,035	101,236	10+
27. Mirzapore	1,071,046	139,893	19+
28. Jaunpur	1,156,254	175,305	15
29 Ghazipur	839,725	109,978	13
30. Ballia	845,418	57,596	7—
Total of Benares Div.	4,809,478	584,008	13

31. Gorakhpur	3,201,180	391,952	12+
32. Basti	1,830,421	280,387	15+
33. Azamgarh	1,492,818	264,615	16
Total of GORAKHPUR DIV.	6,524,419	936,954	14+
34. Naini Tal	323,519	20,071	6
35. Almora	525,104	405	0.077
36. Garhwal	480,167	1,522	0.3
Total of KUMAUN DIV.	1,328,790	21,998	1.65
37. Lucknow..	764,411	79,538	10+
38. Unao	910,915	105,867	12—
39. Rae Bareilly	1,016,864	102,642	10
40. Sitapur	1,138,996	152,714	13+
41. Hardoi	1,121,248	189,301	17—
42. Kheri	959,208	130,802	14—
Total of LUCKNOW DIV.	5,911,642	760,864	13—
43. Fyzabad	1,154,109	174,670	15
44. Gonda	1,412,212	45,148	3
45. Bahraich..	1,047,677	75,590	7
46. Sultanpore	1,048,524	143,937	14—
47. Partabgarh	899,973	110,639	12+
48. Barabanki	1,083,867	91,471	8+
Total of FYZABAD DIV.	6,646,362	641,455	10—
STATES.			
1. Rampur	531,217	43,983	8
2. Terhi	300,819	2,945	1—
Total of STATES ..	832,036	46,928	0.9+
GRAND TOTAL ..	48,014,070	6,083,283	12.66

Note: Chamar Sikhs, numbering 79 males and 39 females, are not included in these tables.

(C) THE DISTRIBUTION OF THE CHAMARS BY DISTRICTS IN THE UNITED PROVINCES, BY RELIGION AND BY SEX

	HINDUS		MOHAMMEDANS		ARYAS	
	CHAMAR		CHAMAR		CHAMAR	
	Males	Females	Males	Females	Males	Females
UNITED PROVINCES	3,098,217	2,977,864	2,759	2,892	829	722
BRITISH TERRITORY	3,073,557	2,955,624	2,740	2,883	829	722
AGRA PROVINCE	2,364,720	2,262,274	2,672	2,834	822	714
MEERUT DIVISION ..	502,083	464,484	89	68	202	117
1. Dehra Dun	13,325	11,155	8	2	2	4
2. Saharanpur	104,949	97,148	66	47	32	26
3. Muzaffarnagar	65,782	58,656	6	8	5	21
4. Meerut	122,082	114,665	18	11	58	..
5. Bulandshahr	94,272	90,589	1	..	36	87
6. Aligarh	101,673	92,271	..	..	69	..
AGRA DIVISION	386,883	328,610	3	1	184	160
7. Muttra	54,510	48,203	..	..	17	27
8. Agra	91,442	82,522	1	..	29	12
9. Farrukhabad	55,436	43,392	..	..	35	28
10. Mainpuri	59,003	46,708	..	..	8	..
11. Etawah	64,186	54,709	2	1	95	93
12. Etah	62,306	53,076	..	..	..	..

SECTION C—(*Continued*)	HINDUS CHAMAR Males	Females	MOHAMMEDANS CHAMAR Males	Females	ARYAS CHAMAR Males	Females
ROHILKHANAD DIVISION	371,303	325,903	36	20	424	419
13. Bareilly	53,234	45,255	1	..	5	11
14. Bijnore	70,649	65,817	5	13	32	28
15. Budaun	80,364	67,632	1	2	21	13
16. Moradabad	94,583	85,612	28	1	366	367
17. Shahjahanpur	54,284	45,771	1	5	..	..
18. Pilibhit	18,189	15,816	..	..	..	..
ALLAHABAD DIVISION ..	354,012	350,803	2,517	2,723	8	12
19. Cawnpore	73,615	64,445	..	..	4	11
20. Fatehpur	36,877	35,676	..	..	..	..
21. Banda	51,641	54,687	..	..	2	..
22. Hamirpur	33,831	35,432	..	..	..	..
23. Allahabad	77,013	80,479	2,517	2,723	..	..
24. Jhansi	46,323	46,034	..	..	2	1
25. Jalaun	34,712	34,050	..	..	..	..
BENARES DIVISION ..	281,259	302,734	3	8	1	3
26. Benares	49,148	52,088	..	..	..	..
27. Mirzapur	68,137	71,756	..	..	..	..
28. Jaunpur	83,459	91,831	3	8	1	3
29. Ghazipur	52,691	57,287	..	..	..	..
30. Ballia	27,824	29,772	..	..	..	..
GORAKHPUR DIVISION ..	457,281	479,662	3	8	..	..
31. Gorakhpur	192,552	199,400	..	..	..	..
32. Basti	137,114	143,273	..	..	..	..
33. Azamgarh	127,615	136,989	3	8	..	..

KUMAUN DIVISION ..	11,899	10,078	11	6	3	1
34. Naini Tal	10,793	9,260	8	6	3	1
35. Almora	250	154	1	..	..	..
36. Garhwal	856	664	2	..	..	..
OUDH	708,837	693,350	68	49	7	8
LUCKNOW DIVISION ..	396,325	364,434	63	41	1	..
37. Lucknow	42,385	37,142	..	11	..	..
38. Unao	55,072	50,795	..	..	..	..
39. Rae Bareilly	49,819	52,823	..	..	..	..
40. Sitapur	79,755	72,945	8	6	..	..
41. Hardoi	100,828	88,465	2	5	1	..
42. Kheri	68,466	62,264	53	19	..	..
FYZABAD DIVISION ..	312,512	328,916	5	8	6	8
43. Fyzabad	84,284	90,386	..	..	..	..
44. Gonda	22,703	22,445	..	..	..	..
45. Bahraich	38,579	36,997	..	..	6	8
46. Sultanpur	69,201	74,736	..	..	..	..
47. Partabgarh	50,916	59,711	4	8	..	..
48. Bara Banki	46,829	44,642	1	..	..	..
NATIVE STATES	24,660	22,240	19	9	..	..
49. Rampur	23,115	20,840	19	9	..	..
50. Tehri Garhwal	1,545	1,400	..	..	..	..

Chamar Sikhs number seventy-nine males and thirty-nine females. *Census Tables, 1911, United Provinces*, Table XIII. pp. 248, 278, 301, 307.

(D) THE DISTRIBUTION OF THE CHAMARS BY DISTRICTS IN SOME PARTS OF THE PUNJAB, BY RELIGION AND BY SEX

	HINDU		SIKH		MOHAMMEDAN	
	Male	Female	Male	Female	Male	Female
PUNJAB	515,969	436,801	97,242	77,908	405	253
BRITISH TERRITORY	414,798	350,355	49,174	39,579	329	169
DELHI DIVISION	213,552	186,020	13,517	10,791	51	8
Hissar	36,533	31,010	2,738	2,386	20	8
Rohtak	25,402	22,125	..	..	8	..
Gurgaon	37,567	35,750	..	..	1	..
Delhi	31,835	28,573	9	..	22	..
Karnal	37,121	31,801	639	552	..	..
Ambala	44,130	36,120	10,076	7,831	..	..
Simla	964	641	58	22	..	..
JULLUNDER DIVISION	153,920	128,521	30,350	25,114	40	26
Kangra	30,325	27,938	239	225	2	1
Hoshiarpur	53,156	44,488	9,327	8,140	20	13
Jullunder	40,295	32,744	4,543	3,645	9	7
Ludhiana	16,163	12,614	12,071	9,871	1	..
Ferozepore	13,981	10,737	4,170	3,233	8	5
LAHORE DIVISION	24,347	18,293	852	355	101	57
Lahore	2,356	1,129	296	68	72	37
Amritsar	668	457	131	68	..	..
Gurdaspur	13,207	10,676	120	86	7	3
Sialkot	4,394	3,507	124	16	2	2
Gujranwala	3,722	2,524	181	117	20	15
RAWALPINDI DIVISION	2,125	1,231	232	146	10	2
MULTAN DIVISION	20,854	16,290	4,223	3,173	127	76
NATIVE STATES	101,171	86,446	48,068	38,329	76	84

Punjab Census Tables, 1911, p. 243

(E) The Distribution of the Chamars by Districts in Certain Parts of Bihar and Orissa, by Religion and by Sex

	CHAMAR—Hindu	
	Male	Female
Bihar and Orissa	520,099	594,368
(1) British Territory	516,189	590,609
Patna Division	125,852	144,967
Patna	29,494	31,467
Gaya	41,857	45,808
Shahabad	54,501	67,692
Tirhut Division	228,240	274,327
Saran	50,066	70,143
Champaran	63,789	69,378
Muzaffarpur	65,054	78,988
Darbhunga	49,331	55,818
Bhagalpore Division	101,238	107,338
Monghyr	27,922	32,375
Bhagalpur	47,284	49,475
Purnea	11,087	10,265
Santhal Paraganas	14,945	15,223
Orissa Division	15,081	17,315
Chota Nagpur Division	45,778	46,662
Native States	3,910	3,759

Bihar and Orissa Census Tables, 1911, p. 101.

(F) The Distribution of the Chamars by Districts in Certain Parts of the Central Provinces and Berar, by Religion and by Sex

	CHAMAR—Hindu	
	Male	Female
Central Provinces and Berar	443,059	458,535
C. P. British Districts	385,704	401,102
Jubbulpore Division	82,679	81,764
Saugor	35,914	35,219
Damoh	21,757	21,562
Jubbulpore	19,999	19,966
Mandla	2,376	2,303
Seoni	2,633	2,714
Narbudda Division	29,849	29,243
Nagpur Division	12,638	11,852
Chattisgarh Division	260,538	278,213
Raipur	98,701	104,377
Bilaspur	103,768	110,553
Drug	58,069	63,313
Berar	16,395	15,789
Feudatory States	40,960	41,644

Census Tables, Central Provinces and Berar, 1911, p. 129.

APPENDIX B.[1]

TANNING, SHOEMAKING, AND LEATHER ARTICLES

The preparation of buffalo, bullock and cow hides, which occupies about a month, consists of two processes, liming and tanning. The hides are soaked, split into sides, and limed. They are left in the pits for from six to eight, or from twelve to fourteen, days according to the season. For each hide one seer (about two pounds) of slaked lime is used and enough water to cover the hide. For every ten seers of lime one of impure soda is added. After three to four, or six to eight, days the skins are removed, and unhaired with a *khurpī*, or scraper. They are then placed in a new lime solution of the same strength as before, but without the soda. When the skins are sufficiently swollen they are taken out and fleshed on a stone slab with a *rāmpī*, or currier's knife. They are then laid in clean water for from four to six hours. Bating (*hāṅgā*) follows. This process is designed to remove the lime and to open the pores so that the hide may be grained and coloured. The first solution consists of ten measures of very old tan liquor and ten seers of the same three times as strong and one seer of *kan*, or rice husk. This is put into earthen vessels and allowed to ferment for about a week. Each vessel holds four sides, which are handled frequently. This process lasts four days. A second bating is done in a solution of water mixed with molasses and *mahwa* flour or with *mahwa* refuse from a distillery. A third bating is then made in a solution the same as the first, except that scraps of fleshing are used in the place of rice husks. The hide is now pliable. It is laid on

[1] See G. H. Walton, *A Monograph on Tanning and Working in Leather*, upon with this section is based.

a slab, scraped on the grained side, and wrung dry. It is then rinsed with old tan liquor, kneaded, rubbed, and wrung dry. Again it is laid in strong tan liquor for from twelve to twenty-four hours, being kneaded and wrung by hand at frequent intervals. The leather is now sewed up with *mūñj* (a grass twine) into a bag, hung up, and filled with tan mixture. This consists of fifteen seers of new and ten seers of half-spent tan bark (*babūl*), water and weak tan liquor. To this mixture are added two to four pounds of small twigs of *baṁḍā*,[1] powdered and mixed with water. The bag is suspended by the neck from a wooden tripod over a *nāṁd* (a large earthen vessel). As the liquor drips through the pores it is poured back into the bag. After twenty-four hours the bag is taken down, the neck is sewn up and the bag is hung up reversed for twelve hours. The hide is then taken down, opened and laid out. It is sprinkled with four ounces of impure salt (*khārī*) and four ounces of bark dust, which are then well rubbed in. The hide is then set out on the grain side with a sleeker. This last, and even the bating, process is often neglected by Chamars. The currying of leather is almost entirely neglected.

Another native process consists chiefly of liming. First, the hides are laid on the floor and roughly fleshed, smeared over with lime-paste and folded up. Each hide is then tied at both ends and placed in a *nāṁd* containing lime solution. The hides are kept in position by means of a large stone. After three days the hides are removed, unfolded and rubbed with lime, after which they are replaced in the *nāṁd* and left for four or five days. They are then taken out, rubbed, scraped, cleaned, and washed with clean water. When the hair and flesh have been completely removed the hides are fit for tanning. The hides, which are now white, are soaked in clean water to which is added a handful of fermented bark-dust paste, and allowed to lie for two nights. The hides are then folded lengthwise and twisted until all the moisture is squeezed out of them. They are then unfolded, wet, and

[1] The mistletoe found on the mango tree.

twisted in the reverse way. This process of wetting and squeezing takes the place of bating. The hides are then treated with tanning materials as above described. After the tanning process has been completed, the leather is curried with salt curds and ghi. This completes the process.

Owing to the excessive use of lime, the leather produced by the Chamar is very porous and of an inferior quality. The tanning is scarcely more than a colouring process. The object of tanning is to produce, by a combination of tannin with the gelatine of the hide or skin, an insoluble, impenetrable substance. The lime destroys to a considerable extent the fibres upon which the tannin acts.

The tanning of sheep and goat skins is almost entirely in the hands of the *Chikwas*, or *Chiks*, Mohammedan leather workers of Chamar origin, who look down with scorn upon the Chamar. This process is, briefly, as follows: The skins, which are received whole from the slaughter-house, flesh outside, are smeared with lime, left for a day, and then turned right-side out. They are then washed and limed, being allowed to lie in the lime for from five to fifteen days, and then washed and fleshed. A thick paste is then made by boiling down *mahua* flour. When it has cooled it is spread over the skins, which are then allowed to stand for eight days; or a gruel of lentil and barley meal and water is prepared, in which the skins are laid for a week, and occasionally handled. The skins are then washed, and laid in tan liquor, being passed from weak to strong solutions in a series of *nāṁds*. This process lasts from eight to fifteen days, during which time the skins are handled two or three times a day, hand-rubbed, and wrung to make them pliable. They are then rubbed with *sajjī* (impure soda) on the flesh side and dried in the sun.

Coloured leathers are made from goat and sheep skins by special processes. To produce red leather *lākh* (lac) is put into the gruel bath. Blue leather is obtained by the use of copper filings, sal ammoniac and lime juice; and black leather by the use of copperas instead of copper filings.

The manufacture of shagreen is in the hands of Mohammedans. The preparation of the skins of various species of deer is as above, except that *sāl* bark is used in the tanning process. *Sāl* gives a rich brown colour, *dhadra* a light yellow, and *babūl* a buff. Combinations of these materials produce shades of colour.

The substances used in tanning are the bark, leaves, and pods of the *babūl* tree *(acacia arabica)*; the *dhadra* or *bakli (anogeissus latifolia)*, a native of the lower Himalayan tract; the bark of the *sāl (shorea robusta)*; *har* and *bahaira*, myrabolams, the fruit of the *terminalia chebula* and *terminalia bellerica* respectively; the bark and berries of the *ghunt (ziziphus xylopyra)*, a jungle tree; the leaves of the *bamḍa*, a parasite commonly found on the mango tree; the fruit, leaves and bark of the *aonla (phyllanthus emblica)*, a tree of moderate size with feathery foliage; the bark of the *amaltās (cassia fistula)*; the leaves and flour of the *mahua (cassia latifolia)*; the bark of the *rhea (acacia leucophlœa)*; the bark of the *avaram (cassia auriculata)*; and the pod of the *dividivi (cæsalpinia coriaria)*. Some materials imported from abroad are also used. *Babūl* is the most valuable tanning agent found in India.

The several kinds of shoes are all made on the same principle. They may be embroidered or otherwise decorated. The shoemaker begins with the sole. A thin piece of leather is smeared with a paste of mustard oil. Over this are laid, first, odd scraps of leather, second, a heavy layer of mud, and third, a thin piece of leather. The curved toe of the shoe forms part of the inside of the sole of leather. The heel-piece is attached in the same way. The maker now puts a couple of stitches of leather thong through the middle of this composite sole to keep it in position for the next step, which consists in stitching on the upper. He begins at the toe, working round with a plain running stitch, boring holes for the thong to pass through. The heel-piece is then trimmed and sewed on to the upper, which is then closed. The toe part is likewise treated. Additional stitching and ornamentation may be added. The commonest kinds of country shoes

17

are called *golpañjā* and *adhaurī.* The latter is generally made for hard work. Other styles of shoes are the *hafti*, something like the English slipper; the *sulemshahī*, a long narrow shoe with a slender *nok*; the *pañjābī*, similar to the former but with characteristic decorations; the *ghetla*, an ugly shoe with an exaggerated curl over the toes, and apparently without a heel; the *gurgābī*, which has no *nok*, made with a buckle over the instep; the *charhāwans*, made of black velvet, with *nok* and heel-piece of shagreen; and the *zerpai*, or half-shoe, with a point and no heel, which is worn by women only.

Among the leather articles manufactured in the villages are thongs; the *maśak*, or water skin, used by the *bihiśtī*; the *kuppā*, a leather jar for holding ghi; the *kuppī*, or *phuleli*, scent bottles; drums, *daṁkā tablā, tāśā* and *dhol*; the *charsā*, *pūr*, or *moth*, usually made of buffalo or of cow hides, and laced in the form of a bag on a circle of wood, and used for drawing water from wells; and *sarnais*, inflated *nīlgāi* hides used to support a cot, and made for working fishing-nets in rivers.

APPENDIX C

BIBLIOGRAPHY

Census Reports.

Imperial and District Gazetteers.

The Peoples of India. Sir H. H. Risley, New Edition, 1915.

Hindu Tribes and Castes. M. A. Sherring, 1872, 1879, 1881.

Punjab Castes. (A reprint of the chapter on "The Races, Castes and Tribes of the People," in the Census of the Punjab published in 1881.) Sir Denzil Ibbetson, 1916.

Tribes and Castes of Bengal. H. H. Risley, 1891.

The Tribes and Castes of the North-Western Provinces and Oudh. W. Crooke, 1896.

Castes and Tribes of Southern India. E. Thurston and K. Rangachari, 1909.

A Glossary of the Tribes and Castes of the Punjab and North-West Frontier Province. H. A. Rose, 1911, 1914.

The Tribes and Castes of the Central Provinces of India. R. V. Russell and Rae Bahadur Hira Lall, 1916.

Brief Review of the Caste System of the North-Western Provinces and Oudh. J. C. Nesfield, 1885.

Memoirs of the Races of the North-Western Provinces of India. 2 Vols. Revised, Sir H. M. Elliot, 1869.

A Rural and Agricultural Glossary for the North-Western Provinces and Oudh. W. Crooke, 1888.

An Ethnographical Handbook of the North-Western Provinces and Oudh. W. Crooke, 1890.

The Origin and Growth of Village Communities in India. B.H. Baden-Powell, 1899.

Hindu Castes and Sects. Bhattacharya, 1896.

History of Caste in India. 2 Vols., S. V. Ketkar, 1909, 1911.

Buddhist India. T. W. Rhys Davids, 1903.

Brahmanism and Hinduism. M. Monier-Williams, 1891.

Bihar Peasant Life. G. A. Grierson, 1885.

Legends of the Punjab. R. C. Temple, 18—.

The Sacred Books of the East. Vols. II., VII., XII., XIV., XXV., XXVI., XXIX., XXX., XXXIII., XLI., XLIII. and XLIV.

The Industrial Organization of an Indian Province. T. Morison, 1906.

Monograph on Trades and Manufactures of Lucknow. William Hoey, C.S.I., 1880.

A Monograph on the Tanning and Working in Leather. H. G. Walton, I.C.S., 1903.

A Dictionary of the Economic Products of India. 6 Vols., G. Watt, 1889-96.

Notes on the Industries of the United Provinces. A. C. Chatterjee, 1908.

Report, Indian Industrial Commission, and Appendix D, 1918.

Hindu Fasts and Feasts. A. C. Mukerjee, 1916.

An Introduction to the Popular Religion and Folklore of Northern India. W. Crooke, 1894.

The Village Gods of South India. Bishop Whitehead, 1916.

An Introduction to the Study of Comparative Religions. F. B. Jevons, 1908.

Primitive Ritual and Belief. E. O. James, 1917.

Fact and Fable in Psychology. Joseph Jastrow, 1901.

Dravidian Gods in Modern Hinduism. W. T. Elmore, 1915.

The Dramas and Dramatic Dances of the Non-European Races. W. Ridgeway, 1915.

Asiatic Studies. First Series. Sir Alfred C. Lyall, 1899.

The Sikh Religion. 6 Vols. M. A. Macauliffe, 1909.

The Religious Sects of the Hindus. Christian Literature Society, 1904.

Kabir and the Kabir Panth. G. H. Westcott, 1907.

The Bijak of Kabir. Translated into English by the Rev. Ahmad Shah, 1917.

Volumes on Rai Das, Maluk Das, Dadu Dayal, and Jag Jiwan Das, in Belvedere Press Series (Hindi), Allahabad.

The Modern Vernacular Literature of Hindustan. G. A. Grierson, 1889.

Reports of the Standing Committee on Mass Movements. United Provinces Representative Council of Missions.

The Mass Movement Commission Report. Wesleyan Mission Provincial Synod, South India, 1918.

Articles in Encyclopædia Britannica. 11th Edition.

Articles in Encyclopædia of Religion and Ethics.

Articles in the Indian Antiquary.

Punjab Notes and Queries.

The Golden Bough. J. A. Frazer, 1913-14.

GLOSSARY

Ajwain—a seed of a plant of the dill species.
Ām—the mango tree, or fruit.
Babūl—a tree (*Acacia arabica*).
Bahlī—a two-wheeled car.
Bahnoī—a sister's husband.
Bairāgī—an ascetic; a devotee; one who has subdued his worldly desires and passions.
Bājrā—a kind of millet; Indian corn.
Bandagī—a mode of salutation.
Bāṁdī lauṁḍī—a maidservant; a bondmaid.
Baniyā—a shopkeeper; a grain seller.
Barāt—the marriage procession.
Batāsā, batāśā—a kind of sweetmeat.
Baṭṭā—the stone roller used with the stone on which spices are ground.
Bel—a tree (*Aegle marmelos*); the fruit of this tree.
Bhagat- a saint; a devotee; a wizard.
Bhajan—a hymn.
Bhaktī--passionate devotion: worship.
Bhangī—a sweeper; a low caste who are scavengers.
Bhelī—a lump of coarse sugar,
Bhūt—a ghost; an evil spirit.
Bichwānī, bichaunī—a go-between; an agent.
Birādarī—a brotherhood; kin.
Bīr—a hero; a powerful demon.
Biyāh—marriage.
Chādar—a sheet; clothing.
Chamrauṭī, chamarwāṛā--the Chamar quarter of a village or town.
Chapātī—a thin cake of unleavened bread.
Chappan—an eastern lid, or cover, for an earthen pot.
Charanāmṛit—water in which the feet of a guru (or other holy person) have been washed
Chārpāī—a bedstead.
Chaudharī—a headman.
Chauk—a square.
Chhappar— a thatched roof; a hut.
Chhatāṁk—the sixteenth part of a seer, or about two ounces avoirdupois.
Chhaṭṭhī—a religious service performed on the sixth day after childbirth.
Chilam—the part of the *huqqa* which contains the tobacco and fire; a clay bowl, with a stem, used for smoking.
Chimṭā—tongs; fire-tongs.
Chirāg—the common earthen lamp.
Chūlha—a cooking-place; a fire-place.
Chuṭiyā—a sacred scalp-lock; a sacred lock of hair.
Dāl—a split pea; pulse.
Dewālī—a fall festival.
Ḍhāk—a tree (*Butea frondosa*).
Dhobī—a washerman.
Ḍhol—a large drum.
Dhotī—a cloth worn round the waist, passed between the legs and fastened behind.
Ḍolā—a form of marriage.
Ḍolī—a kind of sedan for women.
Ḍom—a low, untouchable caste.
Dūb—a kind of grass (*Agrestis linearis*).
Faqīr—a "holy" man; a mendicant.
Gāṁjā—the hemp plant. The fructification when nearly ripe is bruised and smoked for intoxication.
Gāṛī—a cart; a carriage; a coach.
Gaukama—perquisites.
Gaunā—the consummation of marriage; bringing home a wife.
Ghaṛā—an earthen waterpot.
Gobar—dried cow-dung.
Got—kin; family stock; lineage.

Guṛ—Raw sugar.
Guru—a spiritual guide, or teacher.
Haldī—turmeric.
Halwā—a kind of sweetmeat.
Haṁslī—a collar (of metal) worn as a neck ornament.
Haṁsnā khelnā—coarse and obscene jesting.
Holī—a great Hindu festival and saturnalia held at the approach of the vernal equinox.
Hom—a fire-sacrifice; an oblation of clarified butter in fire.
Huqqā-pānī—commensality.
Imlī—the tamarind tree.
Jādū—enchantment; magic; juggling.
Jajmān—hereditary rights; perquisites.
Janwaṁs; *janwās*—the place at the bride's house where the bridegroom and his train are received.
Jawāī—a son-in-law.
Jawār, *Juwār*, *Joār*, *Joar*—Indian corn.
Jhār phuṁknā—to exorcise.
Kachchā—unripe: immature; raw: of imperfect make or texture; clay-built.
Kājal—lampblack.
Kaṁgan, *kaṁgna*—an ornament worn round the wrist.
Kaṁgnā—thread tied around a bridegroom's wrist.
Kanyādān—the bestowing a girl in marriage.
Karāo—a form of marriage.
Kaṛhī—a dish made by boiling meal of pulse with spices and sour milk.
Kauṛī—a small shell used as a coin.
Kharīf—the autumnal harvest.
Khichṛī—a dish made of pulse and rice boiled together.
Khīr—A dish made of rice boiled in milk.
Khūrpī—a scraper; a weeding-knife: a hand hoe for cutting grass.
Kohbar—a place in the house where special preparations are made in connection with marriage.
Korā—an unused pot.
Kuś—a kind of grass (*Poa cynosuroides*).
Laḍḍu—a sweetmeat: a ball of sweetmeat.
Lagan—a notice appointing the day of marriage and other ceremonies connected therewith.
Lauṁḍī—a girl; a slave-girl; a bondmaid; a daughter.
Līpo—("liped")—to plaster with a preparation of cow-dung and mud.
Loṭā—a small metal or earthen pot.
Mahallā—a ward; a street; a quarter of a village or town.
Mahant—a head of a religious order; a religious leader of superior grade.
Mallāh—a boatman.
Māṁḍhā—a marriage shed.
Maṁgnī—betrothal.
Maṭ kor / *Maṭ maṁgra*—a ceremony connected with marriage.
Maṭkorwā—a clay-pit.
Maur—a crown worn by the bridegroom during the marriage ceremony.
Melā—a fair; a religious fair.
Meṁhdī—a certain plant (*Lawsonia inermis*).
Mūsal—a long, heavy, wooden pestle.
Nāch—a dance.
Nāgphānī—the prickly pear (*Cactus ficus Indica*).
Nāṁd—a vat; a large earthen vessel.
Naṭ—the name of a wandering tribe who are generally jugglers and actors.
Nazar—sight; look; fascination; the evil eye.
Neg—presents at marriages and at other festivities made to relations and to particular servants; a fee.
Negī—One who receives a fee, present or *neg*.
Neotā—an invitation.
Neotiyā—a wizard.
Niśānī—sign; token.
Pagṛī—a turban.

Pāṁw pūjā—foot-worship.
Pañch—a council; an assembly of five arbitrators; a leading man.
Pañchak—the first five days of the "light" or of the "dark" moon.
Pañchāyat—a council; a court of arbitration.
Parachhanā—to wave a lamp over the heads of the bride and groom in order to drive away evil spirits; to wave.
Parwānā—a pass, or certificate.
Paṭā—a low stool with four legs.
Paṭrā—a plank to sit upon.
Peṛā—a kind of sweetmeat.
Phera—perambulation; part of the marriage ceremony.
Piṇḍā—a lump, or ball, of rice boiled in milk offered to deceased ancestors.
Pīr—a saint; an old man.
Pīṛhā—a low stool with four legs.
Pitar paksh—the first fortnight of September-October when the Hindus celebrate the customary obsequies to the manes.
Prasād—food that has been offered to an idol, or of which a spiritual teacher has partaken.
Pūrī—a thin cake of meal fried in ghi or oil.
Rabi'—the spring harvest.
Rākṣhas—a demon; a fiend.
Rāṁpī—a shoemaker's, or currier's, knife.
Roṭī—bread; a loaf.
Sādhu—a "holy" man; an ascetic; a mendicant.
Śādī—marriage.
Sāis—a groom; a horsekeeper.
Sagāī—a second marriage.
Śakkar—sugar.
Sālā—a wife's brother: a brother-in-law.
Sāligrām—a sacred stone, commonly found in the Gandak river.
Sant—a devotee; a saint; an initiate.
Sāṛī—a dress consisting of one piece of cloth, worn by Hindu women round the body and passing over the head.
Sarpat—a reed, or reed-grass (*Saccharum procerum*).
Sarsoṁ—a species of mustard.
Satī—a woman who burns herself on her husband's funeral pile.
Sattū—parched grain, such as barley and gram, reduced to meal and made into a paste.
Sauṁf—anise seed.
Sāwan—the fourth Hindu month.
Sayānā—a wizard; a devil-priest.
Sil—a stone on which condiments and other things are ground.
Sinḍūr—red-lead: vermilion.
Sīr, seer—a weight of about two pounds.
Siyānapan—puberty; marriageable age.
Sūgā—a parrot; a marriage pole.
Sūp—a winnowing basket, or fan.
Sūraj—the sun.
Surmā—a collyrium; antimony ground to fine powder.
Sūrya—the sun.
Susrā—a father-in-law; a term of abuse.
Svastika—a magic mark; a symbol of good luck.
Tāj—a crown.
Tālmakhānā—seed of the water-lily (*Anneslea spinosa*).
Tawā—an iron pan on which bread is baked.
Thālī—a tray.
Tījā—the third day of a lunar fortnight.
Ṭīkā—a mark, or marks, made with coloured earths or unguents upon the forehead and below the eyebrows.
Ṭikṭhī—a tepoy or stool.
Ṭonā—a charm; enchantment; magic.
Ubṭan—a paste rubbed on the body before bathing, an "anointing."
Uṛd—a pulse.

INDEX

ADOPTION, 71
Ancestor worship, 114, 115
Animism, 121, 200
Arya, Aryan, 13, 14, 15
Arya Samaj, 237
Ashes, 109, 111, 169

BĀBA FARĪD, 149
Bāmū, 16
Barrenness, 60
Benevolent spirits, 146 ff
Bhagat, 60, 61, 212
Bhairoṁ, 156, 174, 201
Bhīmsen, 153, 173
Bhishma, 153
Bhūmia, 125
Bhūt, 129
Bīr, 133, 170, 192
Birth Customs, Chapter III., 60 ff; abdominal branding, 67; announcement of, 64; bathing, 64, 67; child to breast, 65; *Churel*, 69; cleansing draught, 67; clothing of child, 65; cutting of cord, 63; desires of pregnancy, 61; disposal of placenta and cord, 64; eclipse, 62; eleventh day, 67; evil eye, 67; feast, 68; food for the mother, 65; fourteenth day, 67; incense, 68; knowledge of sex, 61; marks, 62; mother sits on heels, 63; name giving, 68; name kept secret, 69; offering of a goat, 68; precautions against disease, 66; pre-natal sale of child, 61; protection from evil influence, 61; protection of the lying-in room, 63; protective devices soon after birth, 63; provisions for safe delivery, 62; purificatory rites, 67; seclusion of mother and child, 64 f; *Shastī*, 66, 67; sickness and death and evil influences, 69; sixth day (*Chhaṭṭhī*), 65 f, 160; six months after, 68; *Sohar*, 63; superstitions about irregularities, 69; tenth day, 67; to obtain offspring, 60; twelfth day, 67, 68; twins, 67; use of nails, 69; when birth-pains begin, 62
Blood, 143
Brahmans, 64, 67, 73, 75, 76, 80, 85, 95, 134, 159, 172, 176, 179, 202, 208, 210
Bugaboos, 134

CASTE, affiliation and fissure, 32; mixed castes, 14
Castor oil plant, 144
Cat, 124, 160
Chakaliyan, 31
Chamār, a Hindu, 19; a skinner, 20; Brahmanical traditions of origin, 15; current traditions of origin, 15-17; debts, 59, 224; distribution, 20, 21; eats carrion, 20, 45; economic value, 58; field labourers (see Occupations); increasing in numbers, 21; largely farm labourers, 226; numbers in relation to other castes, 20; numbers in relation to Mussulmans, 21; occupancy rights, 58 f, 224; origin of the caste, 17, 18, 19; overcrowding, 225; recruitment of the caste, 17, 18, 19; tanners (see Tanners); unclean practices, 20
Chamār sub-castes, 17, 18, 21 ff; seven divisions of, 22; Aharwar, 25, 29; Alakgir, 27; Chāmar, 24; Chamkatiya, 26; Dhusiya, 25; Dohar, 24, 30; Dosadh, 26,

32; Jaiswar, 22, 23, 30, 211; Jatiya, 18, 22, 23; Kori, 25; Kuril, 16, 24; minor sub-castes, 26, 27 ff; Mochi, 29, 30, 32; Purbiya, 24; (Rangiya), 26; Satnami, 27, 29 f, 219 ff
Chāmū, 16
Chāmuṇḍā, 140, 156, 175
Charman, *Charma*, 12
Chaṇḍālā, 14, 15
Charmakāra, 13
Chaudarī, 48; investiture, 48, 50, 51, 52
Christianity, and future of the Chamārs, 239, 241 ff; converts to, 238
Churel, 69, 123, 129, 130, 142
Concubinage, 37
Courtyard, shape of, 117
Criminals, 26, 27, 235
Crow, 126, 159, 160, 166, 177

DĀDŪ, 214 f
Dādū Panthīs, 104, 214, 215
Dāno, 123, 133
Dasyu, 14
Death, Chapter V., 99 ff; anklets and bracelets broken, 100; anniversary of, 114; at hour of, 99; at house after, 99, 100; away from home, 106; barring the ghost, 107, 114, 135; burning the body, 103; burial of infants, 69; burial of infants while still alive, 69; burying the body, 104; care of bones and ashes, 103, 111, 113; chief mourner's, precautions, 107; Dādū Panth customs, 105; food for the dead, 108, 109, 110, 111; from smallpox, 106; hearth ashes, 111; Kabīr Panth customs, 105, 106; laying the ghost, 114, 135; measuring the corpse, 101; ninth day, 113; other feasts, 112; *piṇḍas*, 101; *Pitar Paksh*, 113; precautions if body kept overnight, 100, 101; purifying the house, 102; remains cast into river, 103; return after cremation, 106, 107; return of the dead, 109; Satnāmī customs, 106; Śiv Nārāyan customs, 104; tenth day ceremonies, 109, 110; third day ceremonies, 108; when procession starts, 101
Dedication of new house, 117
Demons, 128, 129, 234; and disease, 135, 136; and cattle, 183; and trouble, 141; coercion of, 180; devices for scaring, 142-145; village boundary, 140
Dewālī, 119, 200
Dhāk, 123, 144
Disease demons driven away, 182; transference of, 15, 181, 183, 184; village protected from, 182
Disparity in numbers of the sexes, 45; causes, 44
Divorce, 40
Drums, 13, 28, 56, 77
Dūṇḍ, 133

ECLIPSE, 62, 98, 99
Economic needs, 238, 239
Educational programme, 232
Endogamy, 35; exceptions, 35
Environment and caste, 19
Evil eye, 67, 146, 161 ff; and children, 163; and disease, 163; and things of value, 162; protection from, 164
Exorcism, 15, 129, 181
Exogamy, 35

FAIRIES, 135
Female infanticide, 44
Fetishism, 127, 128
Fiends, 134, 160
Five Saints, 147
Folk remedies, a custom, 178; branding, 184; coercion of disease demons, 180; fever, 179, 180; for cattle, 183, 184; garlands, 184; simple remedies, 177; snake-bite, 178, 179
Food, 45; carrion, 22, 24, 30, 45; commensality, 47; leavings of other castes, 45; of other castes, 47; of Mussulmans, 47;

ordinary food, 45 f; pork, 23, 24, 45; women and men eat separately, 47
Furriers, 56

GAYĀL, 131
Ghāsi Dās, 222
Ghīsa Panthīs, 216, 219
Got, 19, 35; see also Exogamy and Totemism
Gorakh Nāth, 149
Gūgā Pīr, 123, 143, 151, 152, 170, 171
Gurus, 202 ff; poets and gurus, 204; travelling, 202; worship of, 202, 203

HANUMĀN, 125, 154, 188, 193
Hem Rāj, 147
Hereditary rights, 52 ff; as field labourers, 53 f; conditions changing, 54
Hides, 11, 22, 53, 228; ox-hides, 11, 28
Hinduism, 237, 244; and Chamārs of to-day, 240
Holī, 118, 174
House building, 116
House burning, 60
House worship, 115

ILLITERACY—ignorance, 225, 230, 231; and environment, 231, 232; what should be taught, 232, 233
Incense, 55, 142
Indra, 172, 199
Infirmities, 235
Intemperance, 45, 73, 82, 89, 90, 95, 225

JAGJĪWAN DĀS, 220, 221
Jinn, 135
Julāha, 32, 33 f

KABĪR, 204 ff, 206, 212, 219
Kabīr Panth, 105, 204 ff, 207
Kālī, 136, 154
Kālū Bīr, 148, 219
Karma, 200

LĀLGĪR, 216, 218
Leather, 11, 12, 13, 25, 31, 53; articles of, 53, 229; exports of, 227; future of leather manufacture, 226; poor quality of, 226; uses of, 12, 13; see also under *Charma*
Leather-worker, 13, 20, 24
Levirate, 39, 96
Luck, 158, 159

MADAIN, 157
Madigā, 31
Magic, 15, 120, 146; and fascination, 164; amulets, 167; ashes, 169; black, 168 ff; charms, 167; exorcism, 179; folk remedies, 177; for rain, 172; in agriculture, 173 f; in medicine, 166, 167; kinds of, 165; love charms, 168; nature, of 164; public, 171 ff; power obtained, 168; sympathetic, 165, 166; tabus, 171; to prevent hail, 173; well-digging, 176; works both ways, 194; see also Witchcraft
Magic symbols, 144
Mahant, 85, 213
Mahua, 123
Malevolent spirits, see Demons
Malūk Dās, 215 ff
Manu, 14, 15
Marriage, Chapter IV., 72 ff; account of gifts, 94; age at bethothal, 74; anointing, 79, 80, 81, 83, 86; arrangements for betrothal, 36, 72; arrangements for wedding, 74; at night, 85, 87; barber, 72; betrothal, 72, 73; betrothal binding, 36, 74; bride's face washed, 91: bride price, 36; Būṛhā Bāba, 77, 82; consummation, 38, 93; crown, 83; departure of bride, 89, 90; *dolā*, 95; double cakes, 75; early, 37, 38; fasting, 82; feasts, 75, 77, 81, 89, 92; fire-sacrifice, 75, 79, 87, 89, 96, 97; fixing dates, 74, 76; foot washing, 85, 86, 93; foot worship 91,

96; furnishing of pavilion, 78; *gaunā*, 94; go-between, 72; groom's procession (*Barāt*), 81, 85; invitations, 75, 96; *kaṁganā*, 75; *karāo*, 96; *kohbar*, 80, 83, 88, 91; law of, 36; *lagan*, 74, 76; magic earth, 75, 77; mock, 98; *nāch*, 75, 84, 89, 92, 94; occupation and, 35; pavilion, 77, 78, 83, 85, 86, 87, 92, 96; *phera*, 87, 88, 89, 96, 98; pledge in a cup, 73; plowbeam, 78, 79, 89, 98; pranks, 88, 93; preliminary inquiries, 72; presents, 74, 85, 86, 90, 91, 94; *raunā*, 94; *sagāi*, 41; *saut sāl*, 97; singing, 77; special forms, 38; special forms of widow marriage, 96, 97, 98; struggle, or challenge, 85, 91, 92, 93; *sūgā*, 78; survival of marriage by capture, 38; square, 87; tests of strength, 85; throwing rice, 91; use of liquor, 73, 82, 89, 90, 95; use of thread, 83; village boundary, 93; village well, 84; visit to landlord, 89, 92; vulgarity, 86, 89; wave ceremonies, 81, 82, 84, 85, 90, 91, 94; widow marriage, 39; wife chosen locally, 35

Masān, 132

Masānī, 133, 137, 169

Mātās, or mothers, 136, 154, 155, 201

Matriarchate, 41

Medals, 34

Midwife—midwifery, 22, 23, 24, 25, 26, 30, 53, 54 f, 63, 65, 154; perquisites, 54; superstitious practices, 55; unsanitary methods, 44, 54 f

Minor castes that work in leather, 28

Mochi, 20, 29, 30, 32, 33 f; some are Chamārs, 33

Moon godling, 198

Moral training necessary, 232

Mussulmans, 20, 21, 238

NĀG Pañchamī, 118, 119

Name giving, 68

Nānak, 206 f

Nat Bāba, 147

Nature gods, 198

Nona Chamārī, 26, 27, 179, 183, 185

OCCUPATIONS, 16, 17, 53, 56 f—bookbinders, 30; cultivators, 25, 26, 27, 57; day labourers, 26, 57; dealers in hides, 32, 57; grooms, 23, 25, 26; harness and saddle makers, 26, 30; house servants, 26; miscellaneous jobs, 23, 27, 28, 56, 57; not chiefly a tanner and leather worker, 57, 226; seasonal, 58; skinners, 29, 30, 233; weavers, 25, 26, 28, 29, 33

Omens, 159, 160

Opprobrious names, 163

Outcastes, 14, 19, 20

Owl, 126, 159, 160, 168, 177

PADUKĀRA, 13

Pañchāyat, 47 ff; by whom summoned, 50; *Chandharī*, 48 f; composition of, 48; fees, 50; fines, 51; importance of, 52; jurisdiction, 49; organization of, 47; penalties, 51; permanent, 48; procedure, 49; when convened, 50

Physical fitness, 235

Pisāch, 132

Polygamy, 37

Poverty, 224 ff; and *begār*, 224; and excess and vice, 225; and small holdings, 224; and ignorance, 225; and inferior processes, 226; and over-crowding, 225

Pret, 131, 132

Puberty, 70; care of girls, 70, 71; initiation of boys, 70

RĀE DĀS, 30, 207 ff, 212

Rāja Bāsuk, 123, 15

Rākshās, 123, 133

Rāmānanda, 204. 207
Rāmānuja, 204
Rām Rāmīs, 216, 219
Religion, 236; of fear, 236; ignorance of, 236; lack of comfort in, 236; moral outlook, 237; objects sought, 236; see Animism, Demons and Benevolent Spirits

SATNĀMĪ, 27, 29 f, 219 ff
Saut Sāl, 168
Shastī, 66, 67
Shoemaker, 13, 22, 23, 24, 25, 27, 29, 33; increasing, 229
Shoes, 13, 56
Shoe factories, 229
Sītalā Mātā, 124, 136, 137, 138, 139, 201
Ṣiva, 16
Śiv Nārāyans, 104, 211 ff
Smallpox; see *Sītalā Mātā*
Snakes, 123, 124
Snake-bite, 124, 178, 179; person bitten lives on for six months, 179
Snake jewel, 134
Social intercourse, *jus primae noctis*, 43; laxity, 41; lewdness, 43; low ideas of women, 42; *satlok*, 43; sexual irregularities, 41, 42, 43; struggle for a higher position, 47; with other castes, 47
Social standing, 20, 232; and crime, 235; and disgust, 233; and food, 233; and religion, 233, 234; and occupation, 233
Śūdra, 14, 15
Sun godling, 198

TABU, 127, 171
Tanners, 11, 13, 23, 24, 27, 29; where most numerous, 56; decreasing, 228
Tanneries, 228
Tanning, 11; tanning sections of Chamārs lowest, 24
Tattooing, 145
Temples, 201
Tenancy, 27
Tiger, 125
Totemism, 126, 127
Transmigration, 200

UNCLEAN, 13, 14
Underpaid (*Begār*), 55 f
Untouchable, 13, 20, 232

VETAL, 129
Village, Chamar group (*chamrauṭī*), 19, 20; organization, 13; outskirts, 13, 15

WAVE CEREMONIES, 67, 81, 82, 84, 85, 90, 91, 94, 144
Witches, 186; a sacrifice, 193; discovery of, 194; punishment of, 195, 196
Witchcraft, 185 ff; and the eating of human flesh, 185; area of influence, 194; art is anti-religious, 187; a precaution against, 197; lore how transmitted, 192; *mantras*, 188, 192; methods of, 187, 188; *mumiai*, 186; and nakedness, 185; naming of spirits, 188; prophecy, 189; service of, 187; signs of, 185; tests of, 195
Wizard, 69, 186; control many spirits, 192; mental and moral level, 197; other names of, 186; other occupations, 197; powers how obtained, 190; 191; source of power, 190
Worship of birds, 125
Worship of trees, 122, 123
Worship of stones, 121

ZAHRA Pīr, see Gūgā Pīr

Printed at the Wesleyan Mission Press, Mysore City.

CPSIA information can be obtained
at www.ICGtesting.com
Printed in the USA
LVHW080039160223
739598LV00015B/1249